"As a therapist who works with survivors of relational trauma, I know that negative self-talk is one of the biggest hurdles in their healing process. This book is a great resource for the 'whys' behind this pattern, and provides some practical strategies for moving through self-doubt and inner criticism. A must-read for those looking to reclaim their sense of self-worth in their healing journey!"

—**Kaytee Gillis, LCSW-BACS**, author of *Breaking the Cycle*

"This well-written self-help book is easy to read and, more importantly, put to immediate use. The author helps us to clearly understand the origins of some of our most self-defeating thoughts by seamlessly blending science-based information with clear, real-life examples. The best part of the book is that she also provides the reader with a hopeful template for change."

—**Hope Payson, LCSW, LADC**, producer of
Uprooting Addiction: Healing from the Ground Up

"In a friendly, direct, and loving voice, Betsy Holmberg teaches about how our brains limit us when we most need their help. I found her insights exciting, freeing, energizing, and above all, practical. What a joy! I know I will return to this book—the sections I've underlined and tagged—for a refresh and a guide as I take on the next challenge, and the one after that."

—**Joanna Barsh**, best-selling coauthor of
How Remarkable Women Lead

"This book is easy to read, understand, and *so* relatable. Betsy is open, honest, and uses humor throughout the book, which makes reading it a delight. You are not your thoughts, so don't let your default mode network (DMN) run the show. Betsy gives lots of examples and tools to choose from. I really enjoyed this opportunity to learn something new and I hope you do too!"

—**Susan Sullivan**, psychiatric nurse practitioner

"Any woman who has ever thought, 'There's something wrong with me,' needs to read this book. Betsy Holmberg's voice is an encouraging counterbalance to the negative self-talk she explores, helping us judge ourselves less and understand our judgments more. Packed full of immediately applicable anecdotes and exercises, women from all walks of life will feel at home in Holmberg's words."

—**Caroline Fleck, PhD**, licensed psychologist,
    and author of *Validation*

"Betsy Holmberg focuses on an important topic in women's health: how problematic negative self-talk is, and how your inner critic can be at the crux of your suffering. She weaves evidence-based neuroscience as a theoretical framework, while offering practical advice and helpful exercises that can help you quell your inner critic, for good. This resource can help restore your inner peace and promote your journey to healing."

—**Tracy Hutchinson, PhD**, psychotherapist,
    faculty at the College of William and Mary, and
    author of *Adult Children of High-Conflict Parents*

The Neuroscience of Why Women
Are So Hard on Themselves

Unkind Mind

—and How to Quiet Your
Inner Critic for Good

Betsy Holmberg, PhD

New Harbinger Publications, Inc.

Publisher's Note

This publication is designed to provide accurate and authoritative information in regard to the subject matter covered. It is sold with the understanding that the publisher is not engaged in rendering psychological, financial, legal, or other professional services. If expert assistance or counseling is needed, the services of a competent professional should be sought.

NEW HARBINGER PUBLICATIONS is a registered trademark of New Harbinger Publications, Inc.

New Harbinger Publications is an employee-owned company.

Copyright © 2025 by Betsy Holmberg
New Harbinger Publications, Inc.
5720 Shattuck Avenue
Oakland, CA 94609
www.newharbinger.com

All Rights Reserved

Cover design by Amy Shoup

Acquired by Jed Bickman

Edited by Kristi Hein

Library of Congress Cataloging-in-Publication Data on file

Printed in the United States of America

27   26   25

10   9   8   7   6   5   4   3   2   1   First Printing

To Buzzy, with my whole heart

Contents

Introduction 1

Part 1: The Science Behind Your Inner Critic

Chapter 1 Why You Are So Hard on Yourself 7

Chapter 2 What Your Self-Talk Sounds Like 15

Chapter 3 Turning Off Your Self-Talk 31

Part 2: Help! I Can't Get It to Stop!

Chapter 4 The DMN Under Stress 49

Chapter 5 The Tired DMN 65

Chapter 6 The Emotional DMN 71

Part 3: Moving Forward as Your Authentic Self

Chapter 7 Reconnecting with Your True Self 81

Chapter 8 Staying True to Yourself in Relationships 99

Chapter 9 Scrolling and Trolling: Navigating Social Media ... 111

Chapter 10 Healing in Therapy and Beyond 117

Conclusion 125

Acknowledgments 127

Bibliography 129

Introduction

Being a woman is exhausting. We feel judged for how we look, what we say, and what we do. We shape-shift to accommodate and care for the people in our lives, whether they are bosses, parents, spouses, or children. When we feel that we didn't do a good enough job, we judge ourselves harshly, pursuing self-improvement to make sure it doesn't happen again. Our thoughts are an overwhelming stream of to-dos, self-criticisms, existential fears, and more.

Science has advanced to the place where we now know why we are so hard on ourselves. It explains where these thoughts come from, why we have them, and how to dismiss them. The reality is that these thoughts do not well up from some deep place within you, revealing the contents of your soul. They are a survival mechanism, akin to your fight-flight-freeze stress response. Hundreds of thousands of years ago, humans lived in "clans" or groups of 50 to 150 people who worked together to secure food and water, raise children, and protect themselves from predators and other dangers. Back then, thinking about fitting in and adapting to clan norms supported this newly formed, socially dependent lifestyle. Today, these automatic thoughts urge you to fit in, do enough to be valued by others, and more—all in an effort to keep you safe. They are like an overprotective parent who is constantly warning you not to walk there or touch that. Their intention is good, but living with them can drive you nuts.

Identifying these thoughts and understanding their rationale will be an eye-opener. Deep-seated insecurities about your personality, intelligence, appearance, and choices will lose their strength as you realize how much of your internal world, and the internal world of others, is driven by this automatic survival machine. When you start using techniques to stop these thoughts, it will open you up to reconnect with who you really are, underneath the smokescreen of anxiety and self-doubt.

The examples and case studies in this book center around the female experience and are a composite of the many women with whom I've worked. However, the science and techniques apply to all. We all have the same hardware. We can all experience negative, self-critical thoughts. No matter how you identify or where you come from, you may benefit from this healing journey.

The idea that you are not your thoughts is not new. The Buddha said it. Meditation teachers say it. Psychology and wellness people say it. The concept is everywhere. However, I never fully believed it—our minds and thoughts created all of the scientific and technological advances that we have today, didn't they? It took learning about where self-talk comes from, and how it differs from other types of thought, to finally "get" what those masters were saying so long ago.

Perhaps, like me, you too feel skeptical about the idea of separating from your thoughts. Walk with me through the science that explains why. We will explore neuroimaging studies, rat studies, survey results, and more. This work is ongoing and updating every day, as we hone our understanding of the brain and how it works.

This book is split into three parts. In the first, you'll learn why we have this voice and get to know its personality, what it sounds like, and how it affects you. You'll get a chance to identify this voice in yourself, exploring what it cares about and how it functions in your life today. You'll learn real-time, tactical techniques for turning off these thoughts. This is not a one-solution-fits-all situation. There are multiple ways to quiet this voice; some will work better for you than others. Their usefulness will also vary by context and type of thought. These will become your new mental health toolkit.

In part 2, we discuss how self-talk gets worse when things get hard. When you're stressed, overworked, exhausted, or sick, it can be physically impossible to turn that voice off. We'll walk through these difficulties one by one and discuss real-world ways to handle times when the thoughts won't stop.

As you get more separate from this voice, and heal the wounds it has created in your life, you will discover extra space and new possibilities. It is a paradigm shift. *If I haven't been the one talking all these years, then who am I?* Part 3 walks you through several exercises to reconnect you with your true self, the one that's been hiding under your fear and anxiety. We'll explore concrete ways to bring this self into your everyday life. This is not about radical life changes, like divorcing your spouse or moving to a new country. It is about showing up authentically, wherever you are in life, and enjoying how it feels to be fully present, without insecurity and worry at the forefront. Lastly, we will explore how this new knowledge fits into the broader context of mental health treatment, explaining both how therapies work and how scientific knowledge of the inner critic can support you in your broader healing journey.

I recommend keeping a journal as you go, to flesh out and heal all the ways this voice has put you down. If you would like to work on this book in tandem with therapy or other treatment modalities, it can only help. So, buckle up, and let's get learning. I am so excited that you have decided to take this journey.

The Science Behind Your Inner Critic

Part 1

Chapter 1

Why You Are So Hard on Yourself

Omigosh, my guests are here. Oh no! Lauren is dressed so much more casually. Am I overdressed? I feel ridiculous. Ah well, too late now. Okay, Rachel just got here. What am I going to say to her? I hate this. Are they happy to be here? I wish my house didn't smell like this. My stomach is sticking out. I should suck it in. Stand up straight. Ugh, forget it. I'm just with women, right? I need a drink.

Does this type of thinking feel familiar? It does to me. Most of us have a nonstop stream of thoughts that monitor us and how people react to us. What's poignant about this example is that it doesn't reflect a challenging situation or relationship. It's hosting a book club. Our thoughts can turn a potentially fun experience into an anxiety-provoking ordeal. It can be such an unpleasant, painful way to live.

For me, this truth hit home after my husband left me. I survived, but three years after the divorce papers were signed, I was still tormented by self-doubt. I thought, *I am damaged goods. No one will want to date me because they'll think something is wrong with me. Moms won't want to be my friend because I'm different. I can't handle being a single parent. It's too hard. I can't do this.* I felt that my divorce had broken me, and I would never be secure or happy again. My anxiety was high, my self-worth low. The situation felt particularly hopeless because I am a therapist. I know what a therapist would say. I know how they would treat me. And I felt that none of it would stop the nagging, everyday self-criticism that haunted me.

In a fit of desperation, I went into the scientific literature. I had exhausted all other avenues, really. And what I discovered shifted my entire perspective on self-talk. In one revelation, I finally saw my inner critic for who it was and started a journey of healing.

Humans believe that our thoughts originate from us because that's how we experience them. It is one constant stream of consciousness, as we narrate our way through the world. Brain imaging studies prove this wrong. Thoughts don't come from one place: There are two distinct networks responsible for what we think. The first is the *central executive network* (CEN), located in the dorsolateral prefrontal cortex and posterior parietal cortex (see the brain scan images in the free tools at http://www.newharbinger.com/54711). When you focus on something, be it a math problem or the color of the sky, you activate this network. The second is the *default mode network* (DMN), found in the medial prefrontal cortex, posterior cingulate cortex, and angular gyrus. This is your inner monologue—the thoughts you listen to throughout your day, and the ones that drive you nuts when you're trying to sleep.

With two competing thought-generating networks, the brain needs a way to decide which gets your attention. For that, it relies on the salience network (SN), in the anterior insula and dorsal anterior cingulate cortex, which you can imagine as an on/off switch. When you focus, the SN turns on the CEN and actively tunes out other stimuli in your environment so you can concentrate. This is how surgeons can operate for six hours and only when they're done realize that they really, really need to go pee. If there is an alarm or threat, the SN turns off your CEN and lets the DMN take over. It can be a big threat—like when a fire truck goes by and you stop whatever you are doing to look for the fire—or a small threat, like being unable to concentrate on that work email because you're hungry. The SN prioritizes your goals against your safety and bodily needs, helping you walk the line between keeping your body alive and handling everything else you want to accomplish in your day.

Evolutionarily, the CEN is the most recent brain area to develop, making it also the most sophisticated. It helps you plan, think logically, and execute complex tasks, from doing your job to managing a household. When you choose to focus your attention on something, the CEN is activated. It is a linear processor: When the task is completed, it shuts down. It doesn't keep you up in the middle of the night, because once it's done what you asked it to do, it turns off.

Like an old house that kept getting new additions, the human brain built more advanced capabilities on top of older structures. Consider the DMN like the original frame of the house. As touched on in the introduction, for the majority of human existence, we lived in hunter-gatherer clans of up to 150 members. Survival depended on fitting into the clan and abiding by clan norms. The DMN supported us by offering constant input on how we were doing and what we could be doing better. It scanned the environment for norms and guidance on what to look like and how to act, criticizing us when we fell short. It enabled us to live collectively, pooling our resources to improve standards of living for all. It supported a critical stage in our evolution.

And we are all still stuck with it today. It runs automatically and cyclically, with one thought leading to the next, which can often feel like a downward spiral. When we actually solve whatever the DMN has flagged as an issue, whether it is our appearance or popularity, it moves on to something else. It doesn't celebrate successes; it is a radar, scanning for the next threat. The Rolling Stones are right: We "can't get no satisfaction," because our brains aren't wired that way.

The DMN is also wildly distracting. When we get caught up in its thoughts, we lose all sense of time and place. It creates blinders to the world around us. We aren't aware of where we are, who is in front of us, or what we are doing. If we spend lots of time listening to the DMN, we can get to the end of the day and wonder where the time went. Or we drive somewhere and don't remember how we got there. For me, if I ever feel that

I have to reacclimate to my environment (*What time is it? What's going on?*), that's my cue that my DMN just went into overdrive.

The DMN lives in the past and future. It looks to the past to understand what went wrong in an effort to learn, and it imagines the future to make sure that we're prepared and can stay safe. Due to its survivalist nature, it tends to be alarmist and scary, which can then produce stress. If you feel particularly worn out at the end of a day, and nothing that big happened, check to see if you were in your head for a lot of it. Listening to DMN thoughts can produce low-grade chronic stress that drains your energy.

The DMN's automatic nature gives rise to its greatest power: It feels like it comes from you. It feels like you create these thoughts, like you are keeping yourself company throughout the day, when in fact you are listening to a primitive safety network that is ill-equipped to handle the complexity of modern life.

Some people express doubt that this network is automatic. I mean, it really feels like we are talking with ourselves! It becomes clearer when you look at its more banal thoughts. Think back to a time when you've had a song stuck in your head. The way those lyrics kept playing endlessly and there was nothing you could do to stop them. Or when a random memory popped into your head. Yesterday I remembered walking into a conference room in Pittsburgh and turning the lights on because it was dark. On what planet is that a helpful thing for my brain to bring up? These are the times that give us a sense of how automatic the DMN really is.

If you take just one thing away from this book, I hope it is this: *You are not these thoughts.* When you think *I'm a loser* or *I can't do it*, that is not actually what you believe. That is your DMN trying to keep you safe in a world it doesn't understand.

To hear how different the CEN sounds from the DMN, let's imagine them discussing the same topic. Let's say that you are newly single after a five-year relationship. You still dream of being married and having kids, so you start online dating again. As you are swiping through profiles, you see one guy in particular who catches your eye. Your CEN says, "He's cute!

Look at that smile. Swipe right." Your DMN says, "He's too good for me. Look at how fit he is. He's skiing and hiking and traveling. Machu Picchu? I'm a loser compared to him. He doesn't want someone like me. He wants someone super fit and adventurous. I will never find anyone."

These are two different worlds. One is pragmatic and purposeful, while the other is condescending and judgmental. We aren't alarmed by the DMN's vitriol, because we've never known what life sounds like without it. We assume that we are the ones thinking these thoughts, and we label ourselves as having low self-esteem, body image issues, or high anxiety, when in reality those personality characteristics belong to the DMN—not us.

There's a scene in *The Wizard of Oz* where Dorothy and the gang are in the palace, quaking in fear as they listen to the "great and powerful" Oz. Toto pulls back the curtain and reveals the wizard as an average old man not wanting to be found out. So now it's time for us to pull back the curtain on our DMNs and expose them for what they truly are.

EXERCISE: DMN Meet-and-Greet

Take a minute now to listen to the DMN. What does yours talk about? What standards does it uphold? How does it criticize you? What does it tell you that you should do?

There are two ways to do this exercise: Either notice when it is going and tune in, or set a timer on your phone to remind yourself to listen up. Write down what it is saying. Is it talking about physical stuff, like food or bodily discomfort? Is it running through all the things you have to do in your day, or bringing up a random memory from grade school? Is it criticizing you, telling you that you are not good enough, or that you don't fit in?

Now that you have witnessed your DMN in action, take a deep breath, knowing that you did not create those thoughts. They happened to you. How does it feel to know that you are separate from these thoughts? In your journal or notebook, write down your observations.

As you go through this exercise and the others, know that they can cause a wide variety of reactions. Some people feel giddy that the voice that has held them back their entire lives is not real or true. Others feel despondent over how this voice has hampered their well-being. No matter what your response is, it's the first step of healing, and you're not alone. Here are a few examples of people discussing the role their DMN has had in their life:

> *Ever since I was little, I had this fear that I would end up broke. No matter how hard I worked or how much success I had, I felt like I was one wrong step away from homelessness. Spending money made me super anxious, and I have always been so hard on myself for needing to spend money at all. I could never understand how people could take vacations or go out for fancy meals. Now I look at those thoughts and realize they aren't real. My DMN can't process something as complicated as saving for retirement! It's just scaring me and stopping me from doing things that could make me happy. What I should really do is talk to a financial advisor and create a plan for my life. I can't believe I spent all that time letting my DMN stop me from living.*
>
> —CARRIE, forty-six

> *After my boyfriend broke up with me, I felt that there was something wrong with me. I was in my early thirties and still not married. Did I mess up? Should I have been dating someone else? Did this mean I would never have kids? I got in my head, and now I realize that my DMN took over. It took the breakup to mean that I was in danger. I mean, it seriously felt like I was getting thrown out to the wolves. But now I can look at the breakup with a new perspective. There is nothing wrong with me. Lots of people are waiting until they are older to get married. Instead of listening to my DMN, I need to have faith that when the timing is right, it will happen. I feel like a weight has lifted off of me.*
>
> —SHARON, thirty-five

I had a really smart brother growing up. He always did perfectly in school and got all these awards. Compared to him, I felt like "the dumb one." My parents never expected me to do as well as he did. I spent more time hanging out with my friends than studying, because what's the point when you're dumb? I feel deep sadness now. My DMN said that I was "the dumb one," and I believed it. I shaped my life according to it. What if I actually was smart? What if I could have been more successful? I feel like I missed the opportunity to do more, and be more, in life.

—DEBORAH, fifty-two

Overall, it is startling how damaging the DMN can be. It makes us feel bad about ourselves, compels us to change ourselves to fit whatever the acceptable standard is, and stops us from taking risks or expressing who we truly are. It makes us feel sad and anxious. We listen to this voice and believe every word it says because we think that it is us.

Then we critique ourselves for having these thoughts. *I'm such an idiot. I shouldn't be so anxious. I should be grateful! Other people have it so much worse than me. What's wrong with me that I can't shake this bad feeling?* We pursue constant self-help, trying to fix our low self-esteem or lack of confidence, when what we really need is to separate from the Nagging Nelly in our brain.

With this knowledge comes great hope. Taking away the mystery of these thoughts makes them lose their power. We can be gentler with ourselves and others, appreciating that we all have this human hardware. We can recognize when our thoughts are getting ugly and/or harder to stop, and take early steps to prevent ourselves from falling down the rabbit holes of anxiety and depression. Instead of trying to fix ourselves, we can turn to proven ways to shut this system down. In the next chapter, we will explore the personality of the DMN. We'll look at three major themes that play out in the majority of DMN talk so you can easily identify them when you experience them yourself.

Chapter 2

What Your Self-Talk Sounds Like

Survival has many levels, from the banality of what you're having for dinner to the complexity of whether a friend likes you, and the DMN is here for all of it. When it comes to the basics, you hear your DMN prattle on about hunger, aches and pains, tiredness, and bodily functions (*I have to pee. Do I go now or wait until later?*). As an example, my DMN is obsessed with temperature. A typical walk with my dog in the winter sounds like *Omigosh I can't do it. This is too cold. I can't stand winter. Why is there wind!? I just need this to be over.* It was only when I realized where these nasty thoughts were coming from that I could start to put them in a place. I tell my DMN now *You're not dying. You're fine. It's just ten more minutes. We'll be home soon.* Many people's DMNs are obsessed with food: what they want to eat, what they should eat, how much they eat, how they look, and more. The world gives us so many cues that looks matter, it's not surprising that our DMNs translate them into a barrage of food/appearance-related thoughts. Unfortunately, our DMNs are happy to create our very own personalized diet culture in our minds.

Nonphysical DMN thoughts group under three consistent themes. First, it cares about *fitting in,* or doing whatever it takes to stay in the clan so you don't get kicked out. Second, it *frets.* It worries about what can go wrong in the future, thinking it can protect you. Lastly, it talks about *failing.*

It downplays your abilities to stop you from taking risks. Or it tells you that you can't handle whatever is going on in your life.

The first step to freedom is understanding how the DMN affects your life today. As you read through these examples, see which ones resonate. Afterward, there will be a quiz to help you identify your DMN's priorities.

Theme 1, Fitting In: Do I Fit In?

> *She just looked away while I was talking. Does she want to be talking to someone else? Say something interesting so she'll like you. Well, that was TMI. Ugh, I have no social skills. Look at what she's wearing. She looks so cute! I look frumpy. Say something funny. I'm not funny. I can't believe I just said that. That was awkward. Why am I so awkward? She'll never want to talk to me again.*
>
> —DEB, twenty-six

The DMN is hardwired to prioritize acceptance above all. It scans your environment to establish what the clan norms are and how you are supposed to behave. This is why you subconsciously check out what everyone is wearing, how they style their hair, and how much makeup they have on. You catalog people's living arrangements, jobs, hobbies, and vacations. You listen for what people talk about, what makes them laugh, and what makes them turn away. The DMN uses it all to figure out how to be accepted by others.

Then the DMN talks to you about how you could be fitting in better. If everyone is wearing a new style of dress, you could feel that you should get one, too. If everyone goes out all the time, you could feel bad about yourself for wanting to stay home and take a bath. Whenever we feel we *should* do something, or we look around and try to "keep up with the Joneses," we are listening to the DMN.

Unfortunately, this desire to conform causes us to shape-shift. We behave differently among our friends than with our family or coworkers.

We have different energy levels, talk about different things, even hold our bodies differently. Many of us structure our lives to fit in. We become a doctor or lawyer because that is what our parents want. Or we mask our gender or sexual orientation because we know our family or community will not accept us otherwise. The unfortunate byproduct of listening to the DMN is that we lose sight of ourselves. Constantly adjusting to the world around us, we have little room left to figure out what really matters to us or who we want to be. The "safer" we are, the less authentic we become.

> Notably, intolerance we see in others also derives from the DMN. Being part of a clan meant that everyone worked together and helped each other. Some were designated hunters, others took care of the children, and so on. This division of labor enabled the clan to thrive. The DMN saw the in-group as inherently good, and out-groups as inherently bad. People from other clans could wage war on us, hurt us, or steal our land. The DMN developed an innate suspicion of people who looked and acted differently from us.
>
> In modern life, this DMN predilection results in prejudice, bigotry, and polarization. If you come across someone who makes you feel bad because of your differences, know it comes from the most primitive, automatic part of them. They need to understand their DMN, separate themselves from this voice, and form their own views before they can treat you with the acceptance, respect, and love that you deserve. Sadly, many people won't want to: Oppression is held in place because some people do benefit from it.
>
> Acknowledging where hurtful and abusive language and behavior comes from doesn't change the fact that they cause real harm. Nor does it address how oppression has been a part of human society since the dawn of time and is institutionalized into systems and culture. This is to say that these topics are complicated and multifaceted. At the least, hopefully this conversation helps contextualize this behavior.

Theme 2, Fretting: Am I Going to Be Okay?

What time is it? 2 a.m.? Ugh, I can't be up. I won't be able to function tomorrow if I don't sleep. The layoffs are what, a week away? What if I get laid off? Will we have to move? Should I be looking for another job now? What if that just gives them a reason to fire me? Ugh, I have a pain in my stomach. Is it cancer? Am I dying? I can't afford to get it checked out. I need to sleep!

—JENNY, thirty-nine

The DMN spends a lot of time assessing risk—past, present, and future. It rehashes things (*Did I really say that stupid thing at the party? Does she think I am an idiot now?*), critiques your every move (*I've let John play video games all afternoon. He's going to have no motivation, and it's all my fault*), and stresses about everything that can go wrong (*This presentation is going to go terribly. Everyone will judge me and think I don't deserve to be here. I am a fraud*).

For many of us, the worry extends past ourselves. We're worried about our parents and our children. We worry about gun violence, drugs, the economy, and more. We worry about the planet, and if climate change has gone too far. If we listen to the DMN and its worries, it can take over a big chunk of our lives.

While worrying can feel productive, like we are doing something to protect ourselves against all the bad things that can happen, it is sadly futile. No one comes out of a worry session feeling like they've cracked the code.

Worry can leave us feeling emotionally and physically drained because the DMN is one of the main controllers of the stress system. Nerve cells link the DMN to the hypothalamus, pituitary, and adrenal network (HPA axis), the seat of the stress system. So when the DMN gives the signal, the

HPA axis launches a full-body stress response, all in an effort to keep you safe. This means that any and all DMN concerns can create stress in the body, hurting us more than helping.

DMN worry isn't just a nuisance. It can impair our performance and prevent us from living up to our full potential. Academic Sian Beilock studied this phenomenon at the University of Chicago. When students put pressure on themselves to do well, they studied more. Studying is good! It strengthens the CEN and SN, helping students focus better. High-pressured students exhibited higher *working memory*—the ability to hold information in their brain and use it to solve the problem. In an exam situation, when the pressure was on, these students performed the same as those with the lowest working memory. The DMN, caught up in worry and pressure, switched off their well-oiled CEN. They couldn't access their learning or reason through complicated problems. Their performance suffered.

Ironically, when a bad event actually happens, we are far more capable of handling it than we anticipate. Reflecting on her terminal cancer diagnosis, poet Andrea Gibson shared, "My whole life I had this terror...this idea that as soon as I got news like this, that I would just be in a cave all curled up and devastated and having no access to joy. And the thing that I've learned through these last two years is...I wasted so much time fearing the emotions that I would have in the future. And that fear that I had in the past is far more than what I'm experiencing right now. The present moment is far more doable than the future or the past."

We are stronger and more capable than we know. We have handled so many big, hard things already and have everything we need in us to handle the things that will happen in the future. We have to stop listening to the DMN, which obsesses over everything that could possibly go wrong, how bad it could get, and how ill-equipped we are to handle it. When we let these worries go, we reduce stress. We feel better and can experience life for what it really is, instead of the DMN's negative idea of what it could be.

Theme 3, Failing: This Is Too Hard

I'm supposed to speak up in meetings, but I just can't. I don't have the experience they all do. I'll look dumb if I ask something I should already know. They're all more senior than me—I'd just be wasting their time. I'll write in my notepad to look like I'm doing something. I hate this.

—SARAH, twenty-five

The DMN believes it is safer to play small than risk doing something large or uncertain. Speaking up in class is scary because people could judge us. Wearing something unconventional could cause people to look at us differently. These can range from micro-nos, like stopping you from calling a friend with whom you haven't spoken in a while, to macro-nos, like talking yourself out of applying to graduate school because your DMN says that you won't be accepted. Unfortunately, these restrictions add up, limiting your life and how you express yourself in it.

The DMN also bets against you. It says that you are not good enough or smart enough. It labels you (calling you stupid, or incapable, or a bad mom—pick your poison) and says you can't handle it. A track record of success doesn't help. People who have high-profile jobs and win big awards still experience the DMN's insecurities. When Michelle Obama entered the White House, she had many accolades under her belt, from attending Princeton and Harvard to her successful corporate law career. Yet in her memoir *Becoming*, she shares, "I'd felt overwhelmed by the pace, unworthy of the glamour, anxious about our children, and uncertain of my purpose." The DMN follows us wherever we go.

When we give these thoughts airtime, the SN learns that we care about them. It feeds us more because it thinks we want more. Over time, this process can produce structural changes in the brain. Think of brain networks like sets of muscles. The ones you use repeatedly are the strongest. Compared to other, less-used pathways, they activate more quickly and run for longer periods without getting tired.

The more we listen to DMN thoughts, the more the brain becomes wired toward them. Over time, it produces hyperconnectivity across the DMN and hypoconnectivity between the SN and CEN, two brain states that are hallmarks of clinical depression. Essentially, in depression, the DMN is in overdrive, while the SN can't engage the CEN and turn it off. We see the world through the negative, self-bashing lens of the DMN. Preliminary evidence suggests that increased DMN functionality may also be seen in anxiety disorders. Experiencing mental health symptoms is not a moral failing or weakness. We need to reframe these as the brain having trouble turning off the self-focused, critical thoughts that affect us all, and to employ methods to rebalance the thought networks instead.

A New Perspective on Suicidality

Suicidal ideation is the extreme of failing thoughts. When things have gotten really hard—because we've suffered a big rejection, or we feel so different from others that we don't see how we fit anymore—the DMN starts saying things like "The world would be better off without you." It looks at all the expectations from the clan, and all the ways that we should be fitting in, and decides that we are not measuring up. Rather than having the clan kick us out, it suggests that we should kick ourselves out instead.

Suicidality still carries a heavy stigma in our society, even though so many people suffer from it. A 2021 US report on adolescent mental health found that one in three girls and one in two LGBTQ+ youth had experienced suicidal ideation in the past year. One possible explanation for why it is common in adolescents is that culturally, our tech-heavy world gives them too many data points that they don't fit in or are failing. As a system, we tend to treat the individual and make assumptions about their stability when we should be looking to their environment to figure out why the DMN is processing so much rejection and uncertainty.

Then there is the aftermath. If a person has suffered from suicidal ideation, they often become afraid of themselves. Their DMN calls them things like unhinged and damaged. Believing these negative profiles, they continue thinking they will never be good enough or capable enough.

Having suicidal thoughts correlates only weakly with attempting suicide. Two-thirds of individuals seen in emergency departments for a serious suicide attempt had only momentary thoughts of suicide, or none at all. Whereas asking about suicidal ideation used to be the gold standard for suicide assessment, several research teams have identified a more predictive model for suicidal risk: suicide crisis syndrome (SCS). This model has two criteria that must be met: first, a feeling that one is trapped and desperately hopeless or urgently needs to escape a life situation that feels inescapable, and second, symptoms including affective disturbance, loss of cognitive control, hyperarousal, and extreme social withdrawal.

Loss of cognitive control appears to be an extreme DMN state, defined as intense or persistent rumination that feels impossible to stop, impairing the person's ability to process information or make a decision. I hope that with more awareness of the DMN and with conversation around understanding and separating from these thoughts, people will be less likely to experience intense DMN takeovers. If they do happen, I hope people will feel more comfortable with seeking help. It is not a personal failure; rather, it is a brain state. For those who still feel worry about their past or present suicidal ideation, know that when the researchers added suicidal ideation into the predictive model, it did not statistically improve the model's ability to identify who would commit suicide and who would not. There is not enough correlation.

It is time for us to change the conversation around depression, anxiety, and suicidality more broadly. They do not reflect an inherent instability in the individual. These thoughts happen automatically as the natural byproduct of an outdated survival system. The person isn't responsible for them; they are being experienced, not created. Knowing this can reduce the heavy burden surrounding these conditions and encourage healing.

If you suffer from depression, anxiety, or suicidality, it does not mean you cannot handle your life. That's the DMN, on a rampage of fitting in, fretting, and failing. There's nothing wrong with you. Your DMN has taken over. Your DMN has been running a marathon, and it's time to get it to sit down. You can break the cycle by separating from this voice so you can see yourself and your life more clearly. Permit yourself to get help. Talk to a therapist, or consider going on medication. The world will look a lot different when you do.

The table here summarizes the three categories of DMN thoughts and how they may play out in your world.

	Primary question	Theme	Sample thoughts
Fitting In	Do I look and act like everyone else?	Belonging	I'm a loser because I don't have a partner. My clothes are ugly. Their life is better than mine.
Fretting	Am I going to be okay?	Physical safety, security	Do I have enough money? Will I get cancer? What if I lose my job?
Failing	Am I good enough?	Worthiness	I'm too dumb. People are polite and just tolerate me. I can't do this.

We end this chapter with a quiz to help you get to know your DMN. What does it care about? How does it put you down? When do you listen to it? How has it affected your life? This work is a first step to separating from your DMN.

Quiz: Your DMN Profile

Some people have a DMN that will be negative about everything. Others find that their DMN is particularly active about one topic in particular, such as appearance or finances. Learning where your DMN spends time helps to identify what areas you'd like to target and heal first.

As you read the following statements, rate how frequently you experience each on a scale of 0 to 7, with 0 not at all and 7 all the time. Don't take too long with any single item—look for an immediate resonance with the ideas.

After you add up and read your results, take out your journal or a notebook and think through the questions posed. Write whatever comes to mind. The purpose is to contextualize your DMN profile. For example, if you score high on appearance and immediately think of your mother's critical comments throughout your childhood, that's a good starting point for understanding how appearance became a survival topic. There is nothing wrong with you or your looks; your DMN learned from your environment to prioritize appearance—or else. This type of reflection can be greatly healing in your journey.

General

_____ When someone talks, I think about what I am going to say back.

_____ I look around and wonder where the day went.

_____ My thoughts keep me up at night.

_____ I stop myself from doing things that could make me look dumb or ugly.

_____ My thoughts spiral to bad places.

Do I Fit In?

_____ I feel like an outsider.

_____ I believe relationships are easier for other people.

_____ I don't feel good enough.

_____ I feel awkward and uncomfortable in social situations.

_____ I look at what other people say, do, and wear, and I copy it.

Am I Going to Be Okay?

_____ I am always imagining the worst-case scenario.

_____ I have a lot of fear (that I will get cancer, or about politics, and so on).

_____ I'm scared of losing my partner/children/friends.

_____ If I don't do everything right, I'll end up miserable and homeless.

_____ I look to the future with dread.

This Is Too Hard

_____ I feel like I am hanging on by a thread.

_____ I struggle a lot in life.

_____ My inner critic says, *You can't do it. This is too much.*

_____ I feel like I'm just making it through.

_____ I am a failure.

Appearance

_____ I critique my body.

_____ I wish I was a different clothing size.

_____ I think other people are so much better looking than me.

_____ I work hard at my appearance.

_____ I believe if I was better looking, I would have a better life.

Relationships

_____ After parties, I rehash everything I said, and I feel embarrassed.

_____ I feel awkward socializing.

_____ Calling a friend makes me anxious.

_____ I feel I waste other people's time.

_____ In conversation, I believe the other person would rather be talking to someone else.

School/Work

_____ I think, *I am not successful.*

_____ At work/school, I worry that people think what I say is dumb.

_____ I constantly imagine getting fired or failing.

_____ I believe my classmates/coworkers are more capable than I am.

_____ I struggle to speak up during classes or meetings.

Finances

_____ I'm anxious about my finances.

_____ I'm scared that I will run out of money and end up homeless.

_____ I don't do self-care because it's too expensive.

_____ I believe I don't deserve to spend money.

_____ Money worries affect my everyday life.

Health

_____ I'm scared that I'm going to get a terminal illness.

_____ I think about my aches and pains a lot.

_____ I worry about not eating the right foods.

_____ I don't trust my body.

_____ I feel that I work hard to not get sick.

Now add up your subtotals for each category. The possible range is 0 to 35.

Low: 0 through 10

Medium: 11 through 20

High: 21 through 35

Low. This is not a category on which your DMN focuses, meaning you do not approach it from a survival mentality. Take a minute to think about this strength. Did someone in your life believe in you, making you feel confident? What skills and capabilities empower you in this category? Flag the positive ways you approach these areas as abilities that can come to bear as you deal with the insecure parts.

Medium. This category sparks some survival instinct in you. You worry, probably adjusting your life to accommodate these concerns. Reflect on how you approach this category. How would you live differently if you felt confident in this area? What actions would you take? How would you feel?

High. Your DMN flags this category as a survival threat. It produces fearful, insecure thoughts in a misguided attempt to keep you safe. This is the perfect area to target using the tools and exercises in the upcoming chapters. Think back to where and when your DMN may have learned these thoughts. Did family members also think this way, or did they share opinions or judgments about you that you took to heart? Did you pick up messages from the media or peers? Were there specific experiences that made these thoughts take hold?

DMN insecurities are contagious. We learn about survival threats from the people around us. For example, if your parents worried about money and felt that nothing ever went their way, you picked up that money is scarce and hard to acquire. If a friend worries constantly about their health, you will wonder about your own. The more we spend time with these thoughts, the more they become woven into the fabric of who we are and how we view the world.

Take a step back and separate from those thoughts. They are not a true representation of the world, how the world perceives you, or even how you feel about yourself. They are the product of an overactive DMN. Take a moment and let that sink in. Write in your journal any observations this idea produces.

Now that you know about the DMN, you can separate from these ingrained fears. You don't have to believe them anymore.

Chapter 3

Turning Off Your Self-Talk

We all want to make the anxious, insecure, and sad thoughts and feelings go away. We actually have found ways to turn it off, without knowing what it is.

Unfortunately, most of these methods can be addictive. They have earned a place in our lives and society because they work at separating us, however briefly, from the constant barrage of internal criticism. Unfortunately, most also have deleterious side effects.

Nicotine, for example, improves performance by decreasing activity in the DMN and increasing activity in the CEN. It not only turns the DMN off but also turns the CEN on, helping us to focus and get things done. When people smoke, they tend to connect with other smokers (as indoor smoking became banned, outdoor smoking groups formed). The social bonding satiates the DMN even more. Put in this context, nicotine addiction makes sense beyond the brain-altering chemical dependency.

Alcohol use is another culturally accepted form of DMN deactivation. When people say that alcohol releases inhibitions, they're recognizing that it impairs the DMN's functional connectivity. When we have a drink, our internal behavior police quiets, so we worry less about what other people think. We feel free to be ourselves.

However, since the DMN makes sure that we stay within socially acceptable norms of behavior, when we drink too much and impair the DMN, we may say or do inappropriate things. The next morning, the DMN reengages, filling us with embarrassment or stress over what we did or said. Over time this cycle makes the DMN stronger. It gets harder to shut it down, making us want a drink even more. Ironic, isn't it? Something that helps us get relief from the DMN actually strengthens it. This science suggests this is a potential pathway toward addiction.

In the early 2000s, I managed the lab of Matt Nock, a brilliant Harvard professor who studied self-harm in adolescents. We ran study after study together, interviewing the teenagers and poring over the data. The results were surprising. From the outside, nonsuicidal self-injury appears to be an out-of-control, outrageous behavior. Yet the logic behind it is surprisingly simple. Participants shared that when they felt stressed and overwhelmed by their thoughts and emotions, harming the body silenced them. Their attention focused on the bodily pain, and the rest of it faded away. Self-harm caused the SN to switch attention from the DMN to the CEN, producing relief.

Many people exercise to "clear their heads," and science backs them up. Children who engage in exercise are better able to pay attention and ignore distractions. Aerobic exercise has been found to increase cortical thickness in the neocortex, strengthening the CEN and SN. But even exercise can have a dark side, with individuals choosing to exercise to such extremes that it leads to exercise addiction, amenorrhea (absence of a menstrual period for three months or more), and even physical injury.

Some choose physically pleasurable activities, like a good meal or a massage. These bring us into our physicality and out of our minds. The DMN shuts down as we focus our attention on how good it all feels. Escapist activities, like watching a movie or reading a book, pull us out of our DMN by focusing our attention on the stories of another. But when we don't connect with a media experience, our minds can wander even more.

Take a minute to think about how you silence negative thoughts. Do you eat a lot of food? Watch hours of television? Get a drink? Or five? Do you go over to a friend's house? Do you get high? Do you go to bed in the middle of the day? Getting a sense of how you have been coping with it up to now helps you flag it in the future. The next time you feel the urge to numb yourself, take a step back and listen to what the DMN is saying. Rather than taking it as a given, write down the thoughts and examine them. What do you notice? What is your DMN scared of? What is it rehashing? Remember, you are *listening* to these thoughts; you didn't create them. Then ask yourself what you would do if you didn't have these thoughts. If you feel you can, go do that thing instead. Or try one of the techniques in the upcoming chapters.

Now to the good part! With a solid understanding of where these negative thoughts come from, we can find methods to stop them that do not make the problem worse. In this chapter, we will explore several techniques for silencing the DMN. We will strengthen the SN so it is better able to shut down the DMN, and exercise the CEN so that focused attention feels more normal and less effortful. These actions will build CEN and SN connections and even spur neuronal growth. Just as a consistent jogging program makes running a mile go from impossible to downright pleasant, we will make tuning out automatic thoughts second nature.

I bet you think I am going to say that you should meditate. The answer is always meditation, right? I have a confession: I don't like traditional meditation. Guided meditations are a blast. Walking and enjoying nature? Sign me up. I love reading meditation books and listening to people who do meditate. But I cringe at the idea of sitting and letting my thoughts wander by. Its presentation as a one-size-fits-all solution irritates me. I don't want to add another task to my already endless to-do list! My DMN loves to pile on, saying that if it only takes fifteen minutes a day to achieve all these amazing health benefits, then I am lazy and weak for not doing it. I am the one to blame for my sadness/unease/pain. My DMN is charming, isn't it?

Brain scans show that meditation *does* shut down the DMN and engage the CEN. That is why it feels effortful—it's hard to work the CEN when you have lived with the DMN on autopilot. But over time, these practices reconfigure the brain. The CEN increases its connections to the DMN. It's as if the CEN has bypassed the SN. It directly and actively shuts the DMN down, even when the person is not meditating. They can be doing other things (like a task in a laboratory) and the DMN will remain quieter than it does in nonmeditating control subjects. Put plainly, meditation strengthens the CEN's ability to maintain focus and weakens the DMN's grip—an effect so profound that researchers call it a *new default mode* wiring of the brain.

But you don't have to meditate to get the effects of meditation. Scientists have delved deeply into the neuroscience of how meditation works, and we can pull out those components to use in our everyday lives.

We can take the root of meditation and build it into how we approach all of life, rather than feel bad about ourselves for not sitting in silence for fifteen minutes a day. The three techniques we will cover—*move the dog*, *ignoring*, and *active listening*—all employ the same intentional focusing as meditation. The only difference is that we can incorporate it into our jobs, relationships, and hobbies.

As you read about these tools, consider using one or all of them. They have no downsides and are complementary, together moving you toward a DMN-free experience.

Move the Dog

Dogs evolved in packs, with a leader, called the alpha, who organizes and leads the group. When it comes to dog training, many approaches focus on how to establish the human as the alpha so the dog will take commands from them. How to do it? In their bestselling book, *Let Dogs Be Dogs: Understanding Canine Nature and Mastering the Art of Living with Your Dog*, the Monks of New Skete and dog-trainer Marc Goldberg write, "We recommend that you walk through the dog's space occasionally. Do it gently, but note his lack of concern as he moves out of your way. The subtle message you're delivering is this: All the space in the world belongs to me, yet I share it with you."

This is exactly the dynamic we want to create with our DMN. We own our attention. We direct where it goes, and we are the ones who choose whether the DMN gets airtime. It's not the other way around.

One option is to anthropomorphize your DMN. Give it a name and imagine its personality. Put it in an outfit or a particular environment. This simple step can offer something to hold onto when you want to separate from your DMN thoughts.

I have found one analogy extremely helpful for breaking me out of a DMN trance and reminding myself that my DMN is not me. It goes like this: I am the CEO of my life, and I have many employees. My best employee is the CEN: It gets sh**t done. I am typically proud of the work that my CEN produces. My senses are all great employees too. I love hearing, seeing, and touching and can engage with the world as a result of their input. But this DMN? A poorly performing employee who sits in my office all day talking gossip and feeling undervalued. No wonder the company is suffering! No one can concentrate. I am the boss, so it is up to me to take charge and send the DMN back to its cubicle. When I close the door on my DMN, I feel peace and relief. I can finally get back to my life.

Some people would rather just call it their DMN. Whatever works for you.

Then, when you catch the DMN in the act, actively focus on something else. This engages the CEN, shutting down the DMN. You know you have succeeded when your mind goes silent. The thoughts stop. Stay in that zone for as long as you can, to train your brain that you do not want to listen to DMN thoughts every day, all day long.

When you're debating what to focus on, think about the Big Three: your body, your thoughts, and your world. First, you can turn inward to experience how your body feels. Check in with your internal sensations. Feel your body pressed against the chair, or the pressure of your feet on the ground. Take a deep breath and be present for the air going in and out of your lungs. Tuning in to your body can be a relief because it gets you out of your head and into your physicality.

If you feel compelled to stay in your thoughts, try cognitive redirection. Become a thought analyzer. What was the DMN just saying? Which thought category (fitting in, fretting, or failing) do these thoughts fall under? If you were a prehistoric person, how would these thoughts have helped you survive? Thank your DMN for its work, and then focus your thoughts on another topic (what's for dinner, how to structure that work presentation).

Or you can focus on the world around you. What do you see, hear, smell, and feel? I like to look out the window and really take in what I see. A tree dappling the sun on the grass, the color of the sky. I intentionally "snap out of it" and pay attention to the world around me. It often feels like I'm waking from a dream. Here I've been in my head when there's a whole world outside. I settle into the "now," and it calms my DMN.

Sometimes we can't settle by just sitting. To get into the present, we have to *do* something in the present. In those moments, try taking a shower, or cleaning the toilet. Take the dog for a walk. Doing something concrete can help you experience the moment, and you've accomplished something to boot. Consider doing an activity in nature: Studies show that being in nature reduces rumination and DMN activation.

There will be times when none of these solutions work. The thoughts can be so scary and real that it feels impossible to disconnect from them. Whether you are worrying about finances, your child's safety, or how you'll cope when someone you love dies, these worries are real and intense. You can't just turn them off. In the next chapter, we will talk through some real situations in which your DMN will not stop, and discuss ways to survive the storm.

For an in-the-moment approach, it can be helpful to do a releasing practice before you try shifting your attention. Get comfortable, whether seated or lying down. Relax your body and breathe deeply. When your breath has steadied and you feel calmer, visualize putting your worries into a balloon. Fill it with everything on your mind. Include all the negative feelings and body sensations associated with those thoughts (heaviness in your chest, heart pounding). Watch it get bigger and bigger. Then, when you've put all of your worries inside it, take a pin, pop the balloon, and see it explode. Or let it go and send it into the sky.

This releasing exercise can also give you the bandwidth to problem-solve your worries using your CEN. For example, if you are worried about finances, do this exercise, then make a plan for how you will address those concerns (your CEN loves planning). Maybe you choose to download a budgeting app or follow a few individuals on social media who give budgeting tips and tricks. Perhaps you browse personal finance books online and choose one to read. These are all positive steps to address your worry. Getting out of your DMN helps you identify creative solutions and long-term actions you can take to feel more secure.

There is real power in identifying your attachment to the DMN and working with it instead of against it. Some people tell me they resist turning off their DMN because they believe it has something valuable to share. They believe its worries are justified, and if they don't worry, they will mess something up or let something fall through the cracks. It feels like you have relied on this voice to succeed up until now, so how can you just let it go?

If this view resonates, try putting your DMN on a schedule. Give it a designated time to run—say, from 5:00 to 5:30 p.m. every day. During the half-hour, take out a journal and write down every DMN thought that comes to mind. Let it run wild. Olympic gold medalist Simone Biles calls this technique her *worry journal*. She says it has greatly helped her manage her anxiety. If you feel anxious at the end of the allotted worry time, try one of the releasing exercises just described, or your favorite relaxation technique. Feel free to share what you write with a trusted friend, or work through it with a professional counselor as well, to further separate you from the DMN whirlwind.

What makes move the dog work is that you customize it to fit you and your life. If you are an action person, focus on actions. If you are a worrier, keep a worry journal. If you are an analyzer, dig into the DMN thoughts and rip them apart. If there is something else you love that hasn't been mentioned here, supplement that activity instead. The more appealing you find the item of your focus, the more likely you will do it.

Ignoring

This technique relies on humans' ability to habituate to nonthreatening stimuli around them. For example, when you live in a city long enough, you barely register the noise. When you are used to working in a busy office, you can block out the distraction of other people wandering around and talking.

This is the SN at its best. When we want to focus on something, it tunes out everything else around it. Unless, of course, something scary happens, at which point the SN will startle you, making you look around and assess whether you need to move or run. Therefore, to experience the DMN as background noise, we need to teach the SN that DMN thoughts are boring. This will be a shift; up until now, the SN has treated DMN thoughts as mission-critical. We can teach it otherwise.

What is something in your life that you easily tune out? Is it your husband's watching the news? Your toddler's music in the car? There are times I love reality TV, and other times when it just sounds like noise. Therefore, when I want to use ignoring, I internally think *There's* The Real Housewives of Me *again*, and I intentionally stop caring.

This method works well when your DMN thoughts are more of a nuisance than a worry—like when it keeps generating random memories, or loops endlessly on what else you could have said in that fight. And every time you intentionally say, *Eh, I don't care*, you teach the SN that DMN thoughts are low priority. You train it to prioritize its focus on something in the here and now.

Active Listening

The previous two techniques apply when you are by yourself and in your head. But a lot of our lives are spent around others! And being with others sure can trigger the DMN.

Think back to its origin: It kept us safe by orienting us to clan norms and aligning our behavior to fit the requirements around us. Therefore, social situations are ground zero for DMN activity.

Am I wearing the right clothes? Do people like me? Am I an outsider? These kinds of thoughts intrude on our interactions with others. They cause us to feel stressed and wish we were cozy in bed watching TV.

Here is a little thought experiment. *Everyone's* DMN is automatic. Therefore, when you worry about whether you are being judged, you are using your DMN to worry about what someone else's DMN is saying. Who knows what their DMN perceives to be the social norms? Who knows what their DMN is prioritizing right now? Perhaps they feel insecure because they are struggling in a romantic relationship or worried about their health, and they bring that insecurity and fear into their conversation with you. Perhaps their DMN constantly talks about fitting in, so when you see the other person look at what you are wearing, they are building up their own catalog of clan-appropriate clothes to wear. Does that have anything to do with you?

People aren't really thinking about us; their DMNs are laser-focused on them and their own survival. We perceive their judgment—based on the tone of their voice, their facial expressions, or words—because our DMN believes that any judgment it can find will teach us how to survive. We feel insecure and unworthy, when it is likely we are picking up on the negative DMN machinations of those around us. And if we work hard to mold ourselves for others' approval, we don't even know if we are getting it right. How do we know that *our* DMN correctly picked up what's important to *their* DMN? We can drive ourselves nuts trying to control the automatic, negative, fear-based thoughts of those around us. It is an exercise in futility.

We need a technique to stop the DMN from running when we're out in the world. We need a way to soothe our nervous system in social interactions so we don't walk away beating ourselves up over what we said or how we acted. This brings us to *active listening*.

For many decades, active listening has been taught as a core skill set for various professions, from education and social work to public administration and sales. It is meant to bring the listener into the present moment so they can bring their full attention to the other person. Think of your friends who are great listeners. You feel that you can tell them anything, and you leave the conversation feeling better about yourself and your situation. They are nonjudgmental, calm, and patient. They make you feel loved and seen.

Then think of your friends who are terrible listeners. Who butt in and tell you how they had something similar happen to them. Or who scare you with fear-based questions or suggestions, thinking that they are helping when in fact they just sent you into a downward spiral. Or maybe they are distracted, spending the whole time looking around to see who else is in the room.

The difference is whether the listener's DMN is at play. When you listen with your DMN, you are half there. You hear the other person, but you're also thinking about yourself. The DMN worries about saying the right thing, runs through similar situations you've had, or even just talks about your grocery list, making it so you don't even hear what the other person is saying. When we socialize from the DMN, we can come off as nervous, self-centered, or dismissive. We can be reactive and critical.

In active listening, you listen from your CEN. You give the other person your full attention, and your inner monologue is quiet. You are fully present.

Active listening is nonjudgmental because you've shut down the part of your brain that judges. You are in an open, loving space. This is the type of attention we want to give the people in our lives. This is us when we are at our best. Studies show that active listening improves the interaction's outcomes, such as satisfaction with the conversation and increased comprehension of the material discussed. Good listeners are seen as more socially attractive, friendly, and understanding.

While teaching methods vary, here are a few good guidelines. First, take a deep breath and move the dog. Give yourself a moment to snap out of your DMN thoughts and become aware of the person in front of you. Second, use the person as your focus point. Feel their presence. Are they uptight? Resigned? Sad? Happy? What are their body position and facial expression? Third, pay attention to what they are saying. Focus on their speech, while engaging with supportive language ("Totally!" "I hear you." "Yeah."). These little words show that you are listening and validate the person's experience and feelings. Fourth, when it is your turn to speak, ask questions that draw out the other person. What did you think about that? How did you feel when that happened? What happened next? While they sound trite, these steps all communicate that you see and hear the other person.

Sometimes this process can be really fun, such as when your friend is sharing a funny story and you get to laugh. Other times, it is rewarding, such as when you get to be there for another in an honest, hard moment.

When people first consider active listening, a common response is, "But I need to think about what I'm going to say! I don't want to leave the conversation hanging!" To combat that fear, have a few questions ready to go so that you can focus on staying in the moment. If it's a social conversation, you could use, "What else is going on with you?" or "How is your family?" If it is work or school, you could introduce the next agenda item or paraphrase the next steps for the group. Or respond with your facial expression and gestures. Whether it is showing empathy with a hug, or giving a nod of comprehension at work, these nonverbal communications can do more than words ever could.

Also, don't underestimate the power of the pause, or taking a second to reflect before you respond. It gives the other person a moment to see if anything else comes to mind that they want to share. It gives you a moment to reflect on what they just said and think about a response. If anything, taking a pause shows that you care. You are sitting with their words and being thoughtful about a response. That is a gift in our hurried, distracted world.

We often overestimate how uncomfortable a pause will be for the person with whom we are conversing. And if it is uncomfortable, remember that it is uncomfortable only for their DMN. Their DMN may sound an alarm: "I don't have anything to say! I'm not good enough! I don't fit in! I stink!" But having the confidence to pause will help calm their DMN down too. This is a safe space. Everyone can take a second to think. We are not performance artists; we are people having a conversation. When you own the messiness of that, your friends will see that they can too.

You may find that as you use this technique in the world, you run across some people with whom you have rich, wonderful conversations, and others with whom you still don't feel safe or comfortable. That's okay. The DMN literature has been really helpful for me to let go of trying to be liked by those people. I used to try to live up to my own made-up standard of what I felt those conversations had to be, to impress the other person, to make them like me, to feel accepted, and more. Now, many times, I find that I don't care. If I don't get that person's approval, or if they don't like me, it's not a big deal. I imagine them in their own DMN cyclone, trying to fit in and survive. Though our DMNs tell us otherwise, we do not need every person's approval to live in the world. I also remind myself that they have their struggles; I just don't know what they are. They may seem like they have won the DMN competition. They may be beautiful or wealthy, or have a gorgeous husband, or live in a beautiful house, or have an amazing job or full social calendar. But their DMN is still hard at work telling them how crappy they are and/or bringing up their own survival fears. No one is immune. I can let them go, stop caring so much, and focus on the interactions that do feel loving and authentic.

With all this in mind, I find that my social anxiety is far less frequent and severe than it used to be. It's not totally gone, but it's way better. And when I do start churning on some silly thing I said to someone I don't know very well, I recognize that my DMN has taken over, and I shift my focus to something else.

Those are the major building blocks for stopping your DMN. As you can see, they all share common roots with meditation. For example, move the dog parallels the meditation practice of focusing on the breath, a mantra, or the word *om*. When instructors teach their students to let their thoughts float by without paying attention to them, they are using the ignoring technique. Active listening is a form of mindfulness, or being fully aware and present.

Now let's take what you have learned and build your own customized, negative self-talk strategy. This is not a one-size-fits-all wellness program. It is an opportunity to use cutting-edge neuroscience to benefit you and your goals for your life. It will be a process of trial and error, as you get to know what you like, what you don't, what feels good to you, and what you can't stand. In the free online tools that accompany this book, at http://www.newharbinger.com/54711, you will find a worksheet where you can map this journey.

To start, go slow. When you train for a marathon, you don't go out and run 26.2 miles on the first day. You start with a walk or a small jog. The same goes for brain training. You want to start with easier steps and build your capacity slowly. Therefore, try to switch out of DMN mode when you feel good. You've had enough sleep, the coffee has sunk in, and there is peace and quiet around you. Test out the feeling when your thoughts are mild (*Ugh, did I just say that?*), rather than in a crisis situation. Try to stop the DMN a few times a day, and see how it feels.

Play around with the process and see what you notice. What focus points help you disengage from the DMN? Which ones don't work at all? For example, some people prefer listening to music, others read Bible verses, and still others find a walk around the block works. Finding your hooks will make you successful. The more tricks you identify, the better your toolbox for turning off the DMN.

If you prefer ignoring, try out different ways to conceptualize your thoughts. Some people love imagining a DMN persona—an annoying roommate, a child who won't stop talking, or a reality TV cast member in a confessional. Others prefer zoning out, like when you're driving on the highway.

Once you find a few techniques that work reliably well for you, practice tuning out the DMN more regularly. Incorporate these into your daily life: waiting at a red light, taking a shower, having lunch—you name it. Make them part of your daily routine.

Know that every moment you disconnect from your DMN tells your SN that you don't care to listen to these thoughts anymore. The more data points the SN receives along these lines, the less likely it will be to prioritize DMN thoughts in the future. You are also strengthening your CEN and its ability to shut down the DMN. You will be able to focus for longer and longer periods.

This is the ultimate irony. Do you want to feel smarter? Be able to retain more information, or focus better at work? *Stop thinking.* Stop letting your DMN drive the car that is you, choosing where you go and cluttering you with its garbage. Letting your mind go silent helps with all of that. I repeatedly ask myself, *What is required of me at this moment?* Often the answer is *nothing*. Doing dishes doesn't require an inner monologue. Walking doesn't require analysis. Some times in my life have required a lot from me—whether studying for a test or building an analytical model at work, I have been overtaxed. I know you've been, too. I step back and relish the fact that right now there is no need to think. It helps me to dismiss whatever my DMN is saying.

Over time, as you do this more and more, you will find that your DMN thoughts get more banal. As you teach your DMN that its derogatory, self-critical nonsense is no longer meaningful to you, your brain will stop delivering it. The DMN will still run—we are stuck with this thing, after all—but it won't hold the power over you that it once did. Instead of kicking you all the time for what you've said or done, the DMN will say things like "I have a headache," "I want a brownie," or "Will Erin be at the dog park? I should text her."

Then, when you do have a self-critical thought, you will be highly aware. *There's my DMN again! Beating me up for what I said.* It's one of those things that once you see it, you can't unsee it. It's hard to give those thoughts as much power when you know they are produced by a brain network that should be wearing animal skins and living in a cave.

Unfortunately, even when you have this level of awareness of your DMN, there will be times you cannot stop it. No matter what you do, it won't shut off. Then your DMN will happily talk about how weak you are because you can't shut it down, or how stupid these methods are because they aren't working.

When I first learned all this stuff, I was so excited. I thought I would walk around smiling like an enlightened monk. Fat chance. I got stressed and had one bad night's sleep, and my DMN came back with a fervor. I couldn't stop it no matter how I tried. So I had to go back to the drawing board to learn why I couldn't shut it down.

What is going on in the brain when we're stressed, tired, or emotional that makes the DMN so strong?

In the next part, we will explore those questions and more. We'll talk about other strategies you can use to get through and feel better. You'll understand why you are working at a deficit when you're stressed, tired, or emotional, so you can be less hard on yourself. You can stop telling yourself there's something wrong with you. I hope this information convinces you that taking care of yourself—including your physical, mental, and emotional well-being—is priority number one, because otherwise the DMN wins. It takes over. When you are driven by the DMN, you become less present and more reactive. Everything gets harder. Let's learn how we can care for ourselves even when that does happen.

Help! I Can't Get It to Stop!

Part 2

Do you ever feel that wellness programs are destined to fail? You read a book and get excited about the promise of well-being. A week later you get a bout of food poisoning, or have a difficult conversation with your spouse, and suddenly none of it seems to stick. Your mood tanks, and it feels like wellness is for other people. You return to your tried-and-true numbing methods. You feel defeated and demoralized.

Join the club! It is really, really hard to keep wellness concepts alive when life hits you in the face.

When we are stressed, tired, or emotional, the DMN feels impossible to stop. I call these conditions *the three nasties* because they bring out the worst in us. Let's talk through why these states put our DMN front and center, and how to take care of ourselves when that happens.

In my work, I've seen that being aware of the DMN makes the hard times a little less hard. People tend to not sink as low and to bounce back faster. You know how after you've been through a bad time, you will often feel guilt or shame for having thought, felt, or acted as you did? As Cara said, "If I got really stressed, that would stress me out. What is wrong with me that I can't handle my life? Do I need to switch jobs? Get help with the kids? Go to therapy? When I realized that my DMN has a loudspeaker when I am stressed, it helped me let go of whatever it was saying during that time. I don't have to switch anything in my life. I don't need to improve myself in some way. I just need to get through, and it will pass. It's been such a relief."

While we can't stop hard times from happening, we can at least equip ourselves with knowledge and solutions to weather the storm effectively. Let's understand what happens in the brain when times get tough.

Chapter 4

The DMN Under Stress

Stress is the nervous system's response to the risk of death. In our hunter-gatherer days, the stress system was attuned to three main dangers: (1) we could get injured or killed by a predator or an enemy in battle, (2) we could get sick, or (3) we could get kicked out of the clan for not following a norm or expectation. Each of these scenarios triggered the same body response, which is protective in the short term and ravaging in the long term.

Stress activates many interconnected systems and pathways. The central hub of this activity, the HPA axis in the brain, starts a hormonal cascade that pumps cortisol—the stress hormone—through the body. Cortisol increases blood glucose levels and pushes that blood to the muscles, helping you to fight or run away as fast as you can. It shuts down nonessential functions, like reproduction and digestion, so all of your energy can go to staying alive. Then it prompts the immune system to launch an inflammatory response. Scientists hypothesize that the immune system gets called on deck since these situations can lead to injury or exposure to unknown pathogens. Whatever happens, your body is prepped to heal.

The DMN plays a critical role in this process, telling the HPA axis whether to launch a stress response, stop one, or do nothing. While the CEN also has connections to your stress system, they are weaker and less direct than the DMN's. From an evolutionary perspective, it makes sense. The primitive thought network designed to keep you safe works hand in hand with your other protective systems. So when your DMN starts to sense danger, it triggers a physical stress response. Where the DMN goes, the body follows.

For example, let's say that you get fired. The DMN sounds the alarm that your safety is in question. It spins about how you will pay your bills, and whether you can find a new job. These thoughts trigger the HPA axis to launch a stress response, which then activates the immune system.

As we previously discussed, the CEN is the brain area that's least critical for survival, so it is the first to get shut down in an emergency. You can't access your CEN when under stress, impairing your attention, working memory, and more. When stress is chronic, this patterning becomes structural. Interestingly, there are direct connections between the CEN and the DMN. The CEN can tell the DMN to shut up and go jump in a lake. But during stress, the CEN loses this ability. Think of stress as having the opposite effect of meditation. Whereas meditation strengthens our ability to focus, and powers the CEN to shut the DMN down directly, stress weakens our focus and reduces its ability to tamp down DMN thoughts.

This can become a vicious cycle. As the CEN weakens, the DMN strengthens, which can lead to even more frequent and greater stress responses. In a sample of teenagers exposed to stressful stimuli, the ones with stronger connections between the DMN and SN exhibited the most intense stress responses. The more we let the DMN run the show, the more likely we will get stressed, and the more extreme our stress responses become.

How Men and Women Process Stress Differently

Rat studies show that stress processing in men and women may differ at the neurological level. In the *locus coeruleus* (LC), one of the brain's stress centers, neurons fire faster in females than in males. Female rats also have more receptors for stress-related neurotransmitters than male rats, and their neurons don't clear the neurotransmitters from the synapses as quickly as males' do, meaning a stress reaction lasts longer in females than males. Sex hormones also play a role: Testosterone inhibits stress responses, and estrogen sensitizes the stress system. Since estrogen levels rise and fall across the menstrual cycle, women experience more variability in their stress response reactivity. There are some times they may get easily stressed, and others when they may feel less affected. On the whole, these results imply that women may have faster, more intense, and longer-lasting stress responses than men.

Sometimes women encounter men who do not get as stressed about a situation as they do. When this happens, it can make women even more stressed. We perceive it as men's choice, and often when they explain why, it *sounds* like a choice. "She will be fine." "Who cares?" These comments can rile us up, making us resent having to play the part of the worried, stressed person in the team, family, or relationship. While there are no easy answers for how to handle these situations, there is understanding. Male and female brains can work differently. Therefore, if a man in your life appears not to share your stress, know it is possible that they truly aren't feeling it. If it feels appropriate, share this literature with them, so they can appreciate why you are more stressed than they are about the same thing (namely, it's *not* that you're psycho). Talk about how you can manage the situation together, by making a plan and divvying up the tasks. The more we understand and respect each other's individual differences, the more caring, supportive, and productive our interactions can be.

Now, are you ready for a one-two punch? We talked earlier about how stress activates an immune response in the body. An immune response affects how our thought networks operate, too.

When we get sick and our immune system kicks into gear, humans and animals alike engage in *sickness behaviors*. We lose our appetite, get lethargic, and retreat from the group. As with everything else we've discussed in this book, there is an evolutionary method to the madness. These actions conserve energy, provide more resources for the body to fight the infection, and protect others from getting sick. Feeling sad and thinking self-critical, pessimistic thoughts (thank you, DMN) accompany these behaviors.

The brain's immune cells, called glial cells, drive these changes. Normally they go around checking on cells to make sure they are healthy and active, and remove the ones that aren't. During an immune response, they can go a little haywire. They eat more synapses and neurons than they should, release neuroinflammatory chemicals, and interfere with the supply of serotonin and other neurotransmitters. Collectively, these brain changes result in our DMN's running the show and making us feel depressed.

Scientists got to the heart of this process by looking at disease treatments that trigger an immune response. For example, when nondepressed individuals were treated for Hepatitis B, about a third of the group became clinically depressed. They felt guilty, pessimistic, and self-critical and lost pleasure in activities that would normally bring joy. Those who exhibited the strongest inflammatory response experienced the greatest deterioration in mood. The stronger the immune response, the worse the mood.

Then researchers looked at it the other way around. Physically healthy people with major depressive disorder exhibited higher inflammation levels than their nondepressed, physically healthy peers. They also had significantly higher levels of activated glial cells.

It is easy to feel discouraged by this discussion. Stress is such a constant in modern life! However, there is a silver lining. Yes, the DMN assesses threats and activates stress responses, but it also inhibits them. It is the great decider.

When we don't understand the difference in our thought networks, this process normally happens *to* us. We listen to ourselves worry, and we own those worries. We think, *Omigosh! I'm so stressed!* and ruminate over how inadequate we are. But armed with this knowledge, we can consciously decide what is stressful or not, rather than letting our DMNs do it for us.

We can also take active measures to reduce stress in our lives. To this end, we will cover top-down approaches that identify the thoughts and minimize them. For example, we'll talk about what to do when the DMN piles on to an already stressful situation or when the DMN creates the stress itself. We will also workshop your major life stressors, seeing the role the DMN plays in making them worse, and doing an exercise to minimize the DMN's influence in this part of your life.

Then we will explore bottom-up approaches or ways to relax the body to ease a stress response. Hopefully, this list will get you started on trying out some solutions that fit your lifestyle. You can incorporate them into your day when you start spinning on some DMN thoughts and feel your anxiety peak, your nerves flare, your energy dwindle, and your mood tank. Combating stress with a top-down and bottom-up combination can help you get out of this state as quickly and painlessly as possible.

Stop the DMN from Making Things Worse

In so many stressful moments, the DMN makes things worse. For example, a college student studying for a test may feel anxious that she doesn't have enough time to learn everything. She then thinks, *Omigosh, I'm going to fail. I'll never get a good job. I'll have to move back in with my parents. My life will be ruined.* This may sound like a young adult just being overly dramatic, but to her DMN, her life *would* be ruined. Her friends would be in a city, living on their own, building their careers, meeting people, and potentially falling in love, while she stagnates in her parents' basement. To the DMN, being separated from your peer group can feel like a death sentence.

Laboratory studies define rumination as "past-centered negative, unwanted, and persistent thoughts." Rumination is essentially listening to your DMN. Researchers have found that when people ruminate after a stressful event, they exhibit a larger stress response to that event—and larger stress responses to future events. Rumination also lengthens the amount of time it takes to bounce back from a stressful event.

Even if a stressful event is positive, the DMN can cause havoc. A classic study using World Series (baseball) data showed that home teams tended to win early games and lose the more crucial final games. Some people hypothesized that the away teams just upped their game; however, with more fine-grained analyses the researchers found that the losses were due to decreased performance by the home team, not stellar plays by the away team. This effect has also been seen in semi-final and championship series basketball games. To unpack what is happening, think about the stress. The fans show up wanting the team to win. This pressure stresses the players out, so they lose access to their CENs and choke. The pressure is positive—the fans are *cheering* for them. And the DMN *still* turns that attention into a negative that destroys performance. It lets everyone down, including, most importantly, the players themselves.

If you are in a stressful situation and your DMN is making it worse, remember that the stress is making it hard, if not impossible, to access your CEN. Your thoughts are more negative and dramatic than normal. When I'm stressed, I give up trying to stop the DMN. Instead, I work hard to not latch onto any one thought so it won't start running. I actively try to *not* believe what the DMN is saying. It feels a bit like I'm putting my fingers in my ears and saying, "I'm not hearing you, I'm not hearing you," but it works. I find I have a better experience if I don't pay attention to my negative thoughts, because it prevents my DMN from playing out a full opera of negative plot twists and consequences. If I do find myself believing a DMN thought, I acknowledge that the negative spiral I just went down came from believing that thought, and I try to let it go. I have a friend who says, "Cancel, clear, delete!" when she gets into these spirals. Whatever works to knock you out of that state.

Catch a DMN-Driven Stress Response

The DMN can produce stress out of nothing. Have you ever thought about a past argument and gotten all riled up, just imagining what you wish you'd said? This is a DMN takeover.

DMN thoughts are negative spiraling. At first it feels productive, like you are processing something, but the more air time the DMN gets, the deeper and darker your thinking becomes, to the point of triggering physical stress symptoms.

Sometimes, as you look back at a particularly hard time in your life, your DMN catches hold of it and immerses you so deeply in the experience that you feel the same feelings all over again. Or you can anticipate a future event with panic and dread, wondering if you'll be able to handle it. Personally, I have a great fear of how I will cope when my parents pass. We are incredibly close. I know it's coming. If I let my DMN spend any time with this scenario, I will feel teary, tight in my chest, and sick to my stomach. Therefore, I actively force myself away from these thoughts. *No, you don't, DMN. We're not going there.* Before, my psychological training would have told me that I wasn't "facing my fears" or processing emotions that require processing. Now I see that a lot of processing isn't helpful or healthy. It is focusing on the bad side of life rather than the good, and bringing my mood and mental health down with the ship.

To be clear, processing traumatic or difficult events is important and healthy—we will cover exactly why when we discuss therapy in chapter 10. However, sometimes what we consider healthy processing is really thinking the same fearful thoughts over and over, without any healing or resolution. This is rumination, just like in the preceding scenarios, except it is over something we're imagining instead of something in our present. In these instances, it is helpful to look at our past experiences and future worries and let our CEN decide how to address them. I, for example, have made a mental note to appreciate my time with my parents now and seek out opportunities to spend more time with them. I have already adjusted my present life to reflect my values, so any further dwelling on the topic is fruitless. For the DMN-driven things in your life like this that haunt you, try addressing them as proactively as you can, then free yourself to turn your attention away from them when they happen.

Another example of DMN-fabricated stress is stereotype threat. If you are aware that others hold a negative stereotype about you—say, that women are bad at math—it will hurt your performance. This effect is robust no matter who you are; when math-proficient white men were primed with the stereotype that Asians are better at math than they are, on a difficult math test they also performed worse than their (white, male) counterparts who did not receive the priming.

We are so used to negatively judging ourselves that even vague, bizarrely worded negative ideas can mess us up. For example, one study successfully impaired people's performance by having them read this one sentence: "There sometimes are and sometimes are not gender differences in math." Isn't that wild? They tested participants' arousal levels and found that the participants who read the priming statement experienced higher arousal (aka stress) and exhibited worse performance on the assigned math test than those who did not see it. The negative statement cued the DMN to ask, "Am I good enough?", which then produced a stress response. The statement caused DMN thoughts, which caused stress, which impaired the CEN, which degraded performance. It's a whole cascade of nasty.

The DMN can internalize these stereotypes as "clan norms" and mess up our lives. If we believe we must be attractive to others to have value, we will spend lots of time on our appearance. If we believe we aren't smart enough to do certain jobs, we will stop ourselves from pursuing opportunities we may love.

Instead of going down this rabbit hole, we need to catch the DMN in the act and turn our focus away from it, as quickly as possible. First, notice the early signs. Do you feel an immediate pang in your gut? Do you feel your heart sink? Do you get that lethargic, down feeling? That's your cue to shake things up. Second, use your in-the-moment tactic of choice. Focus on something else; redirect your attention to the person talking, take a bathroom break, or whatever suits the situation. When you recognize that it's your DMN talking and you choose not to listen, your body gets the message that the stress isn't real. Third, to help your body recover more fully, add one or more of the upcoming body-oriented approaches.

Workshopping Major Life Stressors

We all have hot-button topics that stress us out. Take Pam, who has shaped her whole life around her fear of getting cancer. She reads extensively on it, eats cancer-fighting foods, and refuses to do things that may increase her cancer risk. Though many in society would praise her for being proactive about her health, she actually feels quite stressed and miserable, because fear is the main driver in these decisions. Is cancer her actual reality? No. Is her DMN spinning on the fear of illness? Yes.

Jenny is stressed about money and retirement. Does she have enough to get by? What if she lives a long time? She feels panicky every time she has to spend money, and berates herself for doing anything she deems frivolous, like buying a dress or going out to dinner. Her financial advisor tells her that she is right on track for a comfortable retirement, but that doesn't stop her from getting a pit in her stomach whenever she has to pay a bill, or help her fall asleep at 2 a.m. when she's worried about losing her job.

These are DMN-driven stressors. Whenever these fears pop up, it helps to recognize that the DMN doesn't know what it's talking about.

> **EXERCISE: Rationale and Response**
>
> Grab a piece of paper or your journal, turn it sideways, and draw three equidistant vertical lines. Give the columns headers: Stressor, DMN Rationale, CEN Response. (You can also use the form in the free tools at http://www.newharbinger.com/54711.) In the left column, list the DMN stressors that haunt your life, with vertical space between each. What are your greatest fears and worries? What gives you the most anxiety?
>
> In the DMN Rationale column, write a survival-based explanation for it. Why is the DMN saying these things? How do these thoughts keep you safe, from the DMN's perspective? Imagine back to when these fears were actually helpful. How would these thoughts have kept you safe as a hunter-gatherer?
>
> In the CEN Response column, write down the broader, more realistic view. When you tap into the CEN's rationality and problem-solving, it can greatly reduce the effect of these stressors.
>
> Now review what you've done. What did you learn? How might you use it to help you manage the stress? Are there one or two CEN ideas that you could put on your to-do list today—like starting a budget, or hanging out with friends who aren't worried about cancer? Every little step counts.

When Pam filled out the DMN Rationale, she saw that her cancer-avoidance tactics mirrored how hunter-gatherers had to be cautious about which plants were safe to eat and which could kill them. In the CEN Response column, Pam wrote that there is an element of chance to cancer: Even super-healthy people can get sick, and there are lifelong smokers who never get cancer. Seeing that overall, her DMN thoughts weren't helping her live a joyful or healthy life, she felt empowered to dismiss them. They were adding to her stress and ruining her mood, which she adroitly pointed out are both considered by some to be risk factors for cancer. She realized that stopping the DMN thoughts while living a little more loosely might actually reduce her risk of cancer. She felt empowered, and resolved to practice the techniques to shut down the DMN.

Jenny saw that her financial worries resembled those of a clan wanting to make sure they had enough food to survive the winter. In her CEN Response, Jenny recognized that she had taken awesome steps to protect her financial future. She saw that even if she did lose her job, she was a good worker and had the skills to secure new employment. She recognized that the worry wasn't helping her—it was just stressing her out, which made her performance at work suffer, which put her job more on the line. "It's like the DMN creates the situation it fears the most!" she shared. She also felt more comfortable shutting it down, seeing it as a nuisance instead of a protector.

Here's a final example. When Carol did this exercise, she laughed and asked, "How about all social situations?" She said walking into a party made her start sweating, breathe shallowly, and feel a lump in her stomach. She wrote "Walking into a party" as the stressor. She shared, "Whenever I see all the people clumped together in little groups, I feel like I don't belong. I don't fit in. No one likes me. I'm not as funny or cute or charming as they are. I don't know what to say. I'm too awkward to be there." We then talked about why the DMN might have that reaction. How does it think it is helping you? She said, "Framed in terms of the DMN's viewpoint, I would say it sees the other people as a clan to fit into. It wants me to be accepted because it thinks I need that to survive, and it's telling me what it thinks I need to do to be accepted." I asked her to flip into CEN mode: How true is what the DMN said? What is a more accurate way to frame the situation? Carol said, "They are all just people, with their own baggage. I bet a lot of them are feeling nervous and also want to be accepted and liked, because we all have DMNs, you know? And I bet they are judging themselves way more than they are judging me." The next time Carol walked into a party, she took a pause and remembered our conversation. She ran through all the CEN thoughts she'd explored during our session. She shared, "You know, instead of getting super nervous, I felt compassion for everyone there. I imagined everyone had an invisible list, with all of their insecurities, worries, and tough stuff going on in their lives, and realized that we all have our stuff. They aren't perfect, and I'm trying to fit in with them. They are dealing with their stuff, and I'm dealing with mine. I felt like I belonged. It really helped."

As an aside, there is solid research to support Carol's observation. One study made people feel embarrassed in several controlled situations: a social blunder, an intellectual failure, and listening to someone describe these in an embarrassing way. Researchers had other people look on. When asked afterward, the embarrassed participants thought that they were judged way more harshly by the bystanders than the bystanders actually judged them. The bystanders took context into account, giving the subjects the benefit of the doubt. They also imagined themselves in the same situation and often felt empathy instead of judgment.

If you experience social anxiety, you may find that after understanding and processing the anxiety, you walk into a party feeling less anxious, but you still don't enjoy it. Whereas you used to blame social anxiety for making parties hard, the reality may be that you just don't like the environment. You'd rather talk to people in small groups or one-on-one. This is when you step more authentically into who you are and how you want to live your life, from a place of power and choice, rather than the DMN and its automatic, fear-based reasoning.

Now we move on to bottom-up approaches that help the body regulate out of a stress response. (A deep dive into each method is beyond the scope of this book; I encourage you to check out other sources that go further into the rationale and practice of these methods.)

Deep Breathing

When you're stressed, you naturally breathe from the upper chest area. If you breathe fully, letting the diaphragm drop on inhalation, it calms the stress system. I never connected with the idea of deep breathing as a stress reliever until I learned the technique described here. The first time I did it, by the tenth breath, I could feel my shoulders loosen and drop, and my posture corrected itself. It felt *good*. I kept wanting to take deep breaths just to feel how nice it felt, and in the process, my stress went away. Here's how:

Take a deep breath and let it out quickly. Notice the muscles in the front of your tummy that contract when you do. Keep those muscles contracted as you take in your next deep breath. You want to keep that wall of muscles engaged and firm. Imagine sending your breath into your lower back. Feel your lower back expand as the muscles open up. Keep going until you feel your shoulders drop. For an added effect, stand next to a wall and sandwich a lacrosse ball or similar-sized ball between you and the wall (no need to focus on breathing from your diaphragm for this—it's hard to do both). Roll it around to massage your upper back muscles to release further tension.

Binaural Beats

This is a relatively new stress management technique, and all it requires of you is listening to music. Binaural beat (BB) music sends a slightly different frequency to each ear, which the brain processes as a third tone. This activity synchronizes your brain waves, effectively calming the brain. Research shows that binaural beats can improve connectivity in the CEN, resulting in improved attentional control, focus, working memory, and more. And because the CEN has inhibitory connections to the stress system, it relaxes you. BB music is a great tool to use when you're stressed but need to concentrate, such as when you're studying or working on a presentation. You may find binaural beats on all major music players and YouTube.

Tapping

The Emotional Freedom Technique, otherwise known as tapping, entails stimulating acupressure points on the body that correspond to the parasympathetic nervous system (the nerves that downregulate a stress response and bring the body back to homeostasis). You can tap, rub, or just apply pressure to these areas of the body. You may reinforce the effect with statements that counteract your self-limiting DMN thoughts, like "I am safe," "I am well prepared," "I love myself," and "I am okay."

Guided Meditation

I love guided meditations because they give you something to focus on. As you follow along with the experience, and listen to the music, you activate the CEN. Most meditations include a relaxation component, such as deep breathing or muscle relaxation, which helps to counteract stress in the body. Even if your DMN races through the whole meditation, at least you've taken a few deep breaths and are (hopefully) a little calmer. There are myriad options on YouTube; test them out and see if any work for you.

Shake It Up

Way back when, stress helped humans escape from dangerous situations, so in a stress state we typically have a lot of unused energy that needs to be released. Think about how the body involuntarily shakes after a traumatic event like a car crash. Or how dogs do a full-body shake after being petted for a while. This motion helps clear out the extra energy and return the system to balance.

The next time you feel extra anxious and agitated, try shaking out your arms, legs, and torso. Put on some music and dance vigorously. Go for a run or walk. Use up that extra energy to reset your body. For more information on shaking specifically, check out "therapeutic tremoring" online.

Probiotics for Immune Support

If you've been dealing with chronic stress and mental health issues, you may also want to target your immune system. As we've discussed, chronic stress taxes this system, potentially leaving you with chronic inflammation and reduced capacity to fight viruses and diseases. These states both cause changes in the brain that strengthen the DMN and lead to depressive and anxiety symptoms. By recalibrating the immune system, you can attenuate or reverse these effects, resulting in improved health and well-being.

Consider taking a probiotic, or eating probiotic and prebiotic-rich foods (prebiotics are the food that probiotics eat, helping them to grow and proliferate). Another method to consider is intermittent fasting, though that's probably the last thing you feel like doing when you're in the middle of a stressful time! But try it when things calm down as a way to boost your immune system for the next time you feel overwhelmed.

Helping the body recover from a stress response is a great step toward stopping your DMN during tough times, but it doesn't work miracles. The DMN can still feel hard to stop. If you have chronic stress, your brain may be wired to prioritize the DMN. So when times are tough, consider the following recommendations.

First, prioritize yourself. Often when we are stressed, we feel that we have to be "on" all the time. We stop doing the things that bring us peace and joy—such as exercising, getting our nails done, or reading a book—because we feel they're extraneous. The science says the opposite: They keep us healthy. They bring the CEN online so we can think more clearly and efficiently. Try to work as many self-enriching activities into your life as you can.

Second, recognize the impermanence of things. Life keeps moving and changing. When you're stressed, it can feel like the current situation is too much and you can't make it. The DMN perceives the current situation as life or death, but it isn't. It *will* end, and you will move on to something else. To affirm this, think back to what you were stressed about ten years ago. Are you stressed about that today? Did you get through? Did it end up affecting your life as badly as you thought it would? This too shall pass.

Third, remember your strength. The DMN thinks about everything that can go wrong, so you must force yourself to remember all the things you do right. Think about a challenging situation in the past. You had the strength and courage to overcome it. You worked hard and/or reached out for help and support. You used your innate strengths and talents. The DMN doesn't want you thinking about those things, because that could encourage you to be courageous. It wants you to play small and safe. But you have been strong in the past and are strong now, too.

Lastly, cut yourself some slack. Often we exacerbate our stress responses because we put too much pressure on ourselves to snap out of it quickly. For women in particular, studies suggest that we may be hardwired to get stressed quickly and intensely, and as you have learned in this section, the downstream effects of stressing out are vast. Let go of the pressure and be gentle with yourself. Keep prioritizing rest and rejuvenation and things that make you feel better.

If you're struggling weeks later with unstoppable DMN thoughts and/or low mood and/or no energy, know that it can take up to a month for the body to fully rebalance itself. You will feel better in time.

Chapter 5

The Tired DMN

Sleep. We feel fantastic if we get enough of it, and lousy if we don't. When we're sleep-deprived, just surviving the day can be a struggle. Normal things begin to seem impossible. We count the hours until we can go back to bed again, often to sleep poorly the next night.

When we can't fall asleep, it is usually because our minds won't stop. We keep trying to be calm and meditative, but then it creeps back up, and suddenly we're rehashing all our worries and stressors for the gazillionth time. When it does this toward bedtime, it activates a stress response in the body, however low-level, that overrides our natural sleep cycle, keeping us awake and alert. Therefore, the DMN plays a starring role at nighttime, and in this chapter, we're going to talk about why it does that and how to override its filibuster.

To start, we can actively turn our DMNs off during the day. Neuro-imaging studies show that people with stronger DMNs (aka those that show greater connectivity during waking hours) take longer to fall asleep, and toss and turn more than those with a weaker DMN. They also achieve less REM sleep, which is critical for learning and emotion processing. A strong DMN is hard to turn off, which then makes good quality sleep really hard to get. This explains why there is such a strong correlation between depression and sleep disorders: Studies show that 75 percent of individuals with depression have difficulty initiating or maintaining sleep, or both.

Not surprisingly, an essential step in the falling-asleep process is the DMN shutting down. As your brain waves slow, the front part of the DMN, the medial prefrontal cortex, stops firing with the back parts, like the posterior cingulate. This is when you experience a blissful break from your inner monologue. When the DMN is strong and active, it impedes this process. Your thoughts are all over the place. You can't get settled. You watch the hours tick by. It's exhausting.

Researchers at Brown University and Georgia Institute of Technology explored this phenomenon from a slightly different angle. You know how it is hard to fall asleep or get quality sleep in a new location? To try to figure out why, they watched people fall asleep in a brain scanner. They discovered that during the first night, the left hemisphere of the DMN remained more active than the rest of the brain. It was quick to alert the rest of the brain when something unexpected happened (a voice in the hall, a hotel door closing), which aroused the person to wakefulness. The more alert the DMN was to the outside world, the longer it took for people to fall asleep. The DMN was keeping them awake and alert when they didn't want to be.

So let's say that you have a lot on your mind, and your DMN kept you up all night. You're now groggy and annoyed. This is just the time the DMN runs even more, telling you that you suck, or your life sucks, or that other people are doing it way better than you are. It's an awful experience that can leave you feeling shaky and uncertain about yourself.

Sleep is a critical recovery process that restores and harmonizes the brain and the body. Researchers call sleep "the brain's garbage disposal system" because it clears away the cellular trash that accumulates from a day of use. During sleep, the fluid that bathes your brain is increased by 60 percent. This fluid clears the byproducts of all the day's neuronal activity, rebalances the neurotransmitter levels between synapses, and flushes out the old fluid. One byproduct it clears is beta-amyloid metabolite, a key driver of neurodegenerative diseases like Alzheimer's and other forms of dementia. When you wake up from a good night's sleep and things don't seem quite as bad, or new ideas immediately pop into your head, it's because that massive cleaning has helped your brain run effectively.

When you don't get adequate sleep, you disrupt this process. Your brain is essentially dirty, creating global deficiencies in brain performance. Neurons have a weaker response to activation, and the firing process takes longer. If your neurons were a person, you would say their name five times before they register that you are talking to them. Then it would take them a while to figure out what you were saying, and even more time to respond to you. Everything gets slower and harder.

Just like with stress, the first capability to go is the most advanced, and short-term sleep deprivation creates impairments in the CEN. All of its functions, including alertness, attention, planning, decision making, judgment, reasoning, insight, speech, and language, get disrupted. Your day becomes a lot more difficult. With the CEN impaired, the DMN gets the microphone. This is not the time to try to move the dog, because there's nowhere for the dog to go.

The more nights our sleep is disrupted, the worse it gets, to the point that even the DMN gets impaired. Individuals with persistent insomnia have decreased structural and functional connectivity within the DMN. When the DMN turns off, we lose the part of ourselves that keeps us in line with clan norms. We struggle to be patient and polite or to care about other people's problems. We are left at the behest of our *even more* primitive, aka emotional, brain. The two higher-order thought networks that regulate us and provide structure to our feelings, the DMN and CEN, are not there to calm us down and explain what is going on. So we cry at the drop of a hat, for no reason, or feel angry out of the blue. Our feelings overwhelm us.

There are several key takeaways. First, be extra forgiving of yourself when you're tired. If you have big emotions, now you know why. It's not that you are a mess, or your life is a mess. There isn't anything fundamentally wrong with you. Your brain is dirty and needs a solid cleaning. Do the bare minimum to get by, and prioritize rest and sleep as much as you can. If you're frustrated that you've had a solid night's sleep but still don't feel good, be patient: It can take several nights of recovery sleep to recalibrate the brain. Keep prioritizing rest and sleep, and you will get there.

Second, if your DMN is being a chatterbox at night, remind yourself that it is a master of catastrophe. A pain in your side during the day becomes stage-four cancer at 2 a.m. The slight mistake you made at work becomes grounds for firing you and ruining your career. The dumb thing you said to your crush becomes a reason they will never ask you out.

While there is no research pointing to the evolutionary advantage of having catastrophic thoughts at night, I believe the study on people falling asleep in an MRI scanner hints at something. For early humans, night was our most vulnerable time. It's dark, we can't see, and many predators hunt at night. The DMN keeps us awake in case of an impending attack. No wonder we became pals with dogs, who could stand guard and bark loudly if anything scary approached. It probably made us feel safe, allowing the DMN to take a break, and supported good rest for all.

Nowadays, the DMN tries to make sense of our instinctual fear of the dark by relating that fear to our lives, dramatizing things that didn't seem so bad in the daytime. Appreciating the root of your nighttime horrors may help you to let them go.

Third, try not to fall into the common traps for falling asleep or trying to get back to sleep. Many people attempt to force themselves to stop thinking about their fears, but this never works. You have to replace the negative thoughts with something else. The catch-22, and why night is such a problem for so many people, is that the part of your brain that helps you think about something else, the CEN, is not working very well. It's tired. Therefore, this can feel really, really difficult.

This is where you have to get creative. Sometimes, counting to ten repeatedly is enough to get you out of a DMN cyclone. Or think about something that is somewhat taxing and somewhat boring at the same time, like rehearsing material you just memorized for school, or structuring a new presentation for work.

Some people use gratitude. Try listing the people, places, or events for which you feel grateful. In a study examining gratitude and well-being in life, researchers found that keeping a daily gratitude journal improved people's length and quality of sleep.

When that doesn't work, you may have to resort to grander measures. Taking a hot bath or shower can help you get to sleep. Our body temperature naturally goes through changes over a twenty-four-hour cycle. In the hours leading up to sleep, it drops by 2 to 3 degrees Fahrenheit to initiate quality sleep. When we take a hot bath or shower, we draw out heat from the body's core by bringing it to the surface. Expelling extra heat cools off the core and supports sleep initiation. Studies show that people who take a hot bath or shower for ten minutes or more can experience improved sleep duration and quality. In fact, a hot bath or shower helps people fall asleep almost as fast as the common pharmaceutical sleep aid Ambien. Whereas Ambien users fell asleep sixteen minutes faster than the control group, people who took a hot bath or shower fell asleep seven minutes faster. If you worry that you will awaken others, try using a heating pad to warm your hands and feet. It most likely isn't as powerful as a hot bath or shower, but could trigger an attenuated version of the same cooling process.

If you are up in the middle of the night and nothing seems to be working, try directing your attention elsewhere. Some people swear by e-readers. However, sometimes it is too hard to focus on a book. Then consider getting out your phone or tablet and watching cute, silly, or uplifting videos. I know the blue light disrupts melatonin, making it harder to fall asleep, but which is better: sitting in the dark in a state of panic over DMN thoughts, or calming down your mind and body with videos that make you smile or laugh? It is so easy to berate yourself for using a phone in the middle of the night, but I believe choosing positivity over DMN drama is a powerful action.

Chapter 6

The Emotional DMN

Emotions are a big part of life. They guide us on our journeys: We are drawn to things that make us feel good and move away from things that feel unpleasant. If running makes you feel uncomfortable and unhappy, you're probably not going to run. If painting makes you feel warm and contented, then painting you should do. We also use our emotions to correct ourselves. If we feel guilt for having said something that made another person feel bad, we think twice before saying something similar to someone else. Generally, emotions aid us in building purposeful lives and rich relationships.

This chapter isn't about those productive feelings. This chapter is about when we lose control over our emotions and scream, shout, or say things we wouldn't otherwise say. These are the outbursts that our DMNs love to rehash years after the fact. We all need a survival guide for handling these times.

The Zones of Regulation, an emotion identification framework used in schools, has completely changed the way I talk about emotions with others. It depicts your emotional state as a thermometer with the temperature levels represented by colors. When you are feeling low and sad, you are in the Blue Zone; when you're happy and engaged, the Green Zone. When you're getting frustrated, you're in the Orange Zone, and when that gives way to screaming and slamming doors, the Red Zone. If it feels like your anger has gotten the best of you, you are definitely in the Red Zone.

Neurologically, when you "lose it," you have had an *amygdala hijack*. The amygdala, one of the brain's emotion-processing centers, assesses threats. If it catalogs a weak threat, it enlists the CEN to double-check that everything is okay. You might startle, look up, feel your heart racing a little, then once you've established it's nothing, take a deep breath and get back to what you were doing. If the amygdala decides that the threat is bad, it wants a quick reaction from you so that you can either defend yourself or run to safety. It doesn't waste time asking the CEN for its opinion. Instead, it takes the CEN, SN, and other nonessential functions offline. It makes you jumpy and ready to move. It wants to run the show and get you to safety. The amygdala has hijacked your brain.

Once, when at my computer, I took a sip out of my stainless-steel (aka not transparent) water bottle and felt something large and soft in my mouth. I spit it onto the desk and saw a stink bug waving its antennae. *Gah!* I jumped out of my seat, spit out the rest of the water, jumped around a bunch, and screamed. Did I have *any* control over that reaction? Heck no. My amygdala was in charge, and it decided that a lot of jumping, screaming, and arm-waving was what I needed to do. My dog looked at me like I was nuts.

When we are in the Red Zone, and our amygdala has taken over, everything looks worse because the DMN has the floor. The DMN and amygdala work hand in hand (fun pair, huh?). Has anyone ever apologized to you after an argument and said "I didn't mean what I said"? I used to think that was an excuse, but there's actual truth to it. When we are emotionally taxed, our DMN is driving and saying cruel, hurtful things. These thoughts (that become spoken words) are automatic. So they didn't *really* mean what they said. It was their DMN talking. Now, does that mean that they don't need to apologize, or that every word should be forgiven? Does that mean we should put up with consistent, ongoing emotional abuse? Of course not. But we can all strive to have a little more empathy for each other when mistakes are made.

Now add the complicating factor of female hormones. When these are fluctuating—whether you're in puberty or menopause, pregnant or premenstrual—you can experience an amygdala hijack over much smaller threats. Suddenly you are screaming about something that two days ago you would have let slide. Or you're crying for no reason. Or hating everyone and everything in your life without knowing why. Then you may feel guilt or shame for how these outbursts make other people feel. It may leave you haunted with DMN thoughts that you are a bad mother/employee/friend/wife/what have you, or that you have no control, or that you ruin every good relationship you have, and so on. This is the emotional climate hormones create, and it can feel like a cyclone.

So how do you handle these times? First, do not beat yourself up for it. The DMN sees this behavior and will spin all sorts of mean things about you. It will try to convince you that you don't fit into the clans of your life because everyone else is doing it so much better than you are. It will try to tell you that you are unstable and that there is something wrong with you. The only thing unstable about you is your estrogen levels.

Women rightly get nervous that if they slip up and get emotional, they will be labeled "crazy." Those cultural messages have been around for thousands of years, harking back to when women were hospitalized for "hysteria" and/or a "wandering uterus." The DMN takes those messages and tells us to keep quiet and smile. If we listen to it, we separate from our actual experience.

Luckily, times have changed, and we can choose truth over hiding. The more authentic you are in your life, the more you show others that they can be authentic too. When your body feels two sizes bigger, you want cookies, and your emotions are all over the place, let your family and friends know. Sharing normalizes the experience, helping them to see that they are not unhinged if or when *they* have those feelings. It helps other family members understand that emotions (and hormones) come and go. It's okay to have big emotions. It's okay to have hard days. You can have them and survive.

If you decide to introduce the Zones of Regulation to your family, team, or other group, you gain a whole new language for talking about emotions with each other. For example, if you start feeling sad, frustrated, or whatever, you share which zone you are in and why. It labels the emotion, bringing your CEN to the table as much as possible (it may be just a flyby wave, but that's better than nothing). Not only is it helpful to bring awareness to how you're feeling, but it also helps others know how to support you. When someone says, "I'm getting into the Red Zone," it communicates that their emotions are escalating and their ability to reason is nil. They need space, and using logic with them would only make things worse. The last thing they need is for someone to look at them like they're out of control. They need acceptance and calm from others around them, not judgment. If someone is in the Blue Zone, it can be really helpful to do active listening and let them talk through what's making them sad. Giving people a language for how to describe emotions and a guide for how to support others during different emotional states will help everyone weather those storms together.

If you use this model with kids, it reveals that you have emotions, too. When introducing the concepts to them, share examples of times when they have been in the Blue Zone or the Red Zone, so they see what each entails. Then, when *you* go into the Red Zone, it won't feel like such a shock. We all go into the Red Zone sometimes. You can also help them build skills around managing emotions by asking things like, "What helps you when you go into the Red Zone?" or "What calms you down when you're upset?" With both of your CENs online, you can come up with examples of things that really work. This helps children see that they are bigger than their emotions, and emotions are passing states to manage.

When you've had an outburst, another effective tool to use is *repair*. As coined by parenting experts, repair is a way of taking responsibility and making amends for mistakes. The beauty of this approach is that when you use repair with your kids (and partner, and friends), you model for them how to maintain authentic, connected relationships. You show them that you make mistakes, so when they make a mistake, they will be less likely to perceive you as a high-and-mighty judge who will be disappointed in them. They will be more likely to bring their problems and issues to you. One of the best things my mother ever said to my brother and me was, "You can never mess up as much as your dad and I did." It took the pressure off of us to be perfect and made us feel that we could share our problems, because they probably did something worse.

Here are the steps to repair:

1. Wait until you have sufficiently cooled off from the outburst. You can't repair if you're still in the midst of a blowup.

2. Join the other person wherever they are. This means sitting on the couch or even the floor if you have to. Be at eye level, and make eye contact. This invites them into the connection.

3. Acknowledge how your emotions made them feel. You can say something like, "So I really lost it there. Was that hard for you?" or "I got very angry right then. How did it make you feel?" With adults, it could be, "Wow, my emotions really got the best of me. I was off the charts. How are you feeling about the conversation?"

4. Listen to what they tell you. This is the golden nugget part, and it has to be authentic. Do all the body language things. Look at them. Tilt your head to show that you're listening. Nod along. Say, "I hear you." Offer some attentive pauses. Give them space to open up. Say therapist things like, "Tell me more about that" and "What else were you feeling?" Draw them out. This gives the person an opportunity to articulate their feelings and be heard. The more you repair with the people in your life, the more normal and natural you make these types of conversations for them.

5. Give a heartfelt apology. "I am so sorry that I got upset." Share the reason behind it, using *I-messages* ("I felt angry" instead of "You made me angry"). Give context on why you hit your breaking point: "I was exhausted, and my body was in pain." "I was hungry, and my blood sugar was super low." "I had a really

bad day at work and was feeling super sensitive." "It has been so loud in this house all day, and my ears couldn't take it anymore. You know when you feel overstimulated? That was me." "I've asked several times for you to take out the garbage, and you say yes but don't do it. I feel hurt, like I am a low priority for you. I get tired and frustrated having to ask over and over."

Humans have an uncanny way of blaming their own bad behavior on the context, and another person's bad behavior on their character. We say, "Wow, she's a bitch. She won't even say hello," but then think *I'm so tired I can't handle anything right now.* When you contextualize your behavior for others, you help them contextualize the behavior of other people around them, too. This is the bedrock for strong, fulfilling relationships.

6. Pause, and let the other person talk. Often they will go into problem-solving mode, to try to ensure the situation doesn't happen again. But don't be surprised if they do the opposite, and immediately move on or change the subject. If they turn their attention away from the conversation to something normal, it means they feel their connection with you is repaired.

7. Hug it out, if appropriate.

That was a lot, right? A primary responsibility of the DMN is to make meaning of all the internal sensations and feelings in our lives. However, clearly there are many times when there is no meaning to be made. You are getting sick, so you start feeling depressed. You are stressed, so inflammation is spreading throughout your body and locking your brain in anxious mode. Your progesterone decided not to show up to work this week, and you cry driving to the grocery store. These are the realities of a normal life, and they can feel like pothole after pothole in your quest for well-being.

The main takeaway is to separate your problems from the DMN, as best you can. Don't believe your DMN's nasty quips when you're tired or stressed. Tell yourself *There goes my DMN.* While this is not foolproof, it's knowledge that can help you recover from the hard times. I used to get past a bad situation and think, *Wow, I really need to think through what just happened and how to improve myself so that it doesn't happen again.* I would ruminate on what my DMN said and think about what to change. Now I just let it all go. Dismiss the thoughts, dismiss the meaning, and when things correct themselves, because you got a good night's sleep or the major stressor is over, enjoy the freshness and well-being of feeling good.

Moving Forward as Your Authentic Self

Part 3

When I learned all of this material about the DMN, I found it incredibly liberating. I watched as the things I'd been telling myself, making up my identity, insecurities, and fears, dissolved into *Oh, there's that old DMN again.* If I labeled myself as introverted, awkward, or sensitive, I knew it was just my DMN explaining my behavior. Was there validity to those labels? Perhaps. Did they really represent who I was, or serve me in any way? Definitely not. I felt empty, like a grocery store whose shelves had been cleaned out before a big storm. What, then, was the substance of me?

In this part, we will delve into this question and more. We will explore discernment—what the voice of your true self sounds like, and how it differs from your DMN. We'll talk about choices, small and large, and how to use them to create a life that resonates with your true self. We'll talk about times when there are no choices, and how to honor and support your true self in situations in which you do not have control. We will then explore how to integrate all of this knowledge with other healing practices, like therapy.

These steps can lead to a shift in your overall orientation. Most of us walk around listening to the DMN talk at us all day, and we structure our lives according to its whims. It says we look out of style, so we browse new clothes online. It says we should have a partner, so we go on date after date, even if we find the process painful and exhausting. It says we need to volunteer to be a good person, and we grudgingly do so. *Eat less. Be more interesting. Have more friends. Make more money. Put your kids in soccer.* The DMN gives us endless directives on how to live a "perfect" life.

Instead, let's explore how to listen to your heart, with the help of your CEN. What makes you feel good? What makes you feel happy? What do *you* really care about? As you let the old expectations, beliefs, and worries go, you can shape a life as unique as you are. A beautiful life that is truly yours.

Chapter 7

Reconnecting with Your True Self

What is my true self? Think of the true self as who you are when you strip away all of your cultural, social, and familial conditioning. This includes the stuff that you're supposed to do and the stuff you have to do. If those boundaries were not in place, what do you really want to do, and how would you want to act? Ask yourself: *If money and social norms were unimportant, how would I spend my time?*

Many people love traveling. I believe it is because it helps them put the DMN in perspective. They see people living happily with completely different customs, rules, and living situations, showing that maybe the DMN doesn't have it all figured out. Maybe there is wiggle room in what is accepted and required.

To connect with the true self, we need to zoom out from the brain and our inner monologue and connect with the body's other brains. Other brains? Yes. As scientists have deepened their understanding of our internal structures and organs, they have discovered clusters of neurons, or brain cells, within the body that are not just extensions of the brain. Operating as independent control centers, they function autonomously from the brain, yet also communicate with the brain. We have two of these "body brains"—in the gut and the heart.

The brain in your gut drives most of the digestive decisions your body makes, from how much acid to secrete to when to move food down the tract. Some researchers conjecture that the mind doesn't want to be bothered with the lower-order processes of digestion. Others believe that digestion is so critical to our health and well-being, from securing energy for our cells to managing immune threats, that an independent brain is required to perform this role well. However you look at it, the gut brain serves a vital role in our physical and emotional well-being.

Imagine you are on stage and about to give a talk. It has been your professional dream to present to a crowd. You've rehearsed your presentation dozens of times, yet you still feel a little sick and hope you can make it through without your voice cracking and people noticing. It's frustrating. You're never going to get promoted if you choke! What's going on here?

The gut brain has bidirectional links to the amygdala, SN, and DMN, meaning it can activate these systems and vice versa. For example, when the DMN decides something is threatening, it signals the amygdala to launch a fear-based stress response. The DMN and amygdala both send signals to your gut: *You're in danger!* Your gut brain registers that something is amiss and produces GI symptoms like cramping, nausea, and more. This is how we get the "butterflies in the stomach" effect and other stress-based gastrointestinal conditions, like irritable bowel syndrome (IBS).

The opposite is also true. When you eat something spoiled and feel awful, your gut alerts your DMN. Your DMN works to identify what you ate that was bad and makes a mental note to avoid it in the future. Now having the floor, the DMN may wonder if this is really serious and if you are going to die. It is happy to keep going, discussing all of the other scary and worrisome things in your life—unless you shut it down voluntarily.

Therefore, when people tell you to trust your gut, it means listening to another survival system in the body, working in tandem with the DMN.

When it starts rumbling and feeling bad, look at the context to determine whether to pay attention or not. Are you walking in a dark alley? Your gut is warning you that the situation could be dangerous. Are you at your first tap-dancing class, nervous about how you'll do or what people will say? Perhaps it's saying you'll be worse than the other students and will be rejected. Perhaps it tells you that you are too old, or that you don't deserve to do something fun, or that you should be doing something more productive. Think of it this way: If the DMN and your gut brain are warning you about real and/or imminent danger, listen to it! If it is on you about anything else (e.g., fitting in, failing, or fretting thoughts), tune it out and listen to your heart brain instead.

The heart has a distinct bundle of neurons attached to it that controls heartbeats, blood pressure, and more, independent of the rest of the brain and nervous system. Imagine a heart transplant. If the heart relied on the brain to oversee these operations, it could not be transplanted from one body to another. But because the heart brain is there managing everything, transplants become possible. Over time, nerve connections grow between the head brain and heart brain, enabling those feedback loops to resume.

Like the gut brain, it connects with the SN, so that it too can direct your brain's attention to important events. However, it also connects with the CEN. In research articles, scientists expressed surprise that the heart would speak directly with the most sophisticated parts of the brain. Why would those areas engage in something as basic as heart rate and blood pressure? Yet seen holistically, it makes complete sense that our hearts drive attention and focus.

Let's say that after work you are meeting up with your roommate and your dog at a park. When you see them, it's as if the rest of the park fades away. You focus on calling and waving to your dog to get him to run toward you. You then pet him and delight in his tail wagging. This is an example of when your heart drives your attention. The connection between the CEN and the heart shows that we are designed to pay attention to, and focus on, the things that we love.

When we let the DMN drive, it cuts us off from our hearts, because there are no pathways connecting them. We can be with the people and animals we love the most, in the places we love the most, yet feel alone, removed, stressed, nervous, sad, and more, all because we are wrapped up in the circular script of a fear-based DMN. Where the DMN is, love can't go. As a result, we can push ourselves to work too hard and ignore our needs for play and connection. We can exercise too hard, not listening to our body's screams for rest and nourishment. We can give too much, dismissing our own wants and needs. When we are in DMN mode, we lose connection to ourselves and others. We lose our ability to love.

When you connect with your heart, you are getting direct information from your true self: what you care about, and what drives you. Leading with the heart is the best way to connect with who you really are and create an authentic, fulfilling life. In the rest of this chapter, we will explore several different ways to activate and engage this vital part of your being.

Step One: Listen to Your Loves

EXERCISE: Loves of My Life

Think back over your life, and identify the things you loved. What lit you up? What felt effortless and joyful? We start with childhood because that is a time when we are less aware of societal and familial expectations. The DMN hasn't even fully developed yet, so it is naturally a time that we live more from the heart.

Did you love horses? Drawing? Reading? Soft fuzzy things? Choreographing dances? I want you to feel into that experience. What could you sit for hours doing or imagining? Write these in your journal or a notebook under the heading "Childhood." (Or use the free tool at http://www.newharbinger.com/54711.)

As you got older, you gained exposure to new experiences and ways of being. You discovered people outside of your community doing things in new and different ways. What excited you then? What did you want to do professionally but stifled because it wouldn't be accepted or pay enough money? What hobbies or interests did you discover? Write these down under the heading "Young Adulthood."

As you kept growing and expanding, you found new loves. What do you love about your life now? Is it your morning coffee? Your time with your friends? A good night's sleep? Is it cooking and eating a delicious meal you made yourself? Spending time with your children? Have you picked up new hobbies? Or even seen someone do something on television and wished that you could do the same? Write down those loves under the heading "Now."

These lists contain the breadcrumb trail that will lead you to your true self. When you feel that you are getting off track or not feeling like yourself, return to these lists. Feel into these loves. Imagine yourself experiencing these loves, and sit in the feeling for a while.

This sensation is who you really are. The feeling of warmth, expansiveness, and lightness reflects your true self. This is why gratitude practices work so well. When people sit in gratitude, they bring their attention to love and therefore into alignment with who they really are. Everything else in their body relaxes: their mind, their emotions, their breath, and their muscles. DMN thoughts magically disappear because the person has directed their focus to the current abundance in their life. Try introducing more gratitude into your day as another tool for connecting to your true self.

Now that you have a sense of how these loves make you feel, I want you to incorporate them into your life as many ways as you can. Take Rebecca, who reconnected with how much she used to enjoy the beach as a child. While on vacation, she randomly walked a beach and felt such joy and peace that she called her mother and said, "I just had the craziest experience. I was on a beach and felt incredible. I stayed for six hours!" Her mother said, "Of course! You were obsessed as a child. I could never pull you away." Rebecca put a beach screensaver on her computer, added a beach painting to her living room, and made a playlist of wave sounds to use as background noise while working. She got involved in beach-related projects, giving small donations to a nonprofit that is cleaning up the ocean, and wearing a Save-the-Ocean bracelet with pride. Surrounding herself with beach-related objects and activities lights up her true self.

Cynthia has always loved dogs. In her twenties now, she told herself she couldn't have a dog because she was single and wanted to meet a partner first. She felt she had to be flexible in her life so she could date, go away on weekends, and so on. She rationalized that it would be unfair to an animal if she stayed out late and was away from home going to parties. When asked if she liked parties and bars, she said, "God no. They are exhausting, and I hate being hungover in the morning. But I'm single! What else am I supposed to do?"

Her DMN had designed her life in the hopes of meeting someone with whom to settle down—by doing activities she hated. The DMN is a real stinker, isn't it? Cynthia realized that anyone with whom she would connect probably wouldn't like that stuff either, and she was unlikely to meet them drunk at a party. She got her dog, goes to bed at 10 p.m., goes on dog-walking dates now, and has never been happier.

Children intuitively understand this. When they meet someone for the first time, they lead with "What's your favorite color? Your favorite animal?" They ask what the person loves as a way to get to know them. If you feel you can be more authentic around children than adults, less guarded and more playful, it may be because kids put their loves front and center in their lives. They don't judge us from a DMN viewpoint, because their DMN hasn't developed yet.

The listing exercise may have reminded you of a hobby or activity you've let go of. Most people let their hobbies fall by the wayside for a variety of DMN reasons: *They're too expensive; I should be saving money for retirement; it's a waste of time when I should be cleaning the house or working.* I used to tell myself similar things until the science opened my eyes. Spending time doing a hobby is how you exercise your CEN! First, the focus—whether it's on sewing stitches or hitting a tennis ball—activates your CEN, reinforcing and strengthening those pathways. Second, since the activity is connected to your heart, it gives you double CEN power. You get all the happy, fun feelings of having done something you enjoy. Then when you do have to return to work, parenting, or other duties, you can use your beefed-up CEN to focus better and for longer. You can more easily ignore your DMN thoughts, keeping you out of anxiety and depressive states that would only drain your energy further. Hobbies are *not* too expensive or a waste of time. They are a key pathway to a productive, authentic, and happy life.

Some people walk through this exercise and feel disheartened instead of excited. You are not alone! There are several reasons why you may find it depleting. First and foremost, you may be exhausted. You have given so much of yourself—to your family, friends, job, and more—that there is nothing left over. You have given and given and given, and your cup is empty. Tired, depleted people don't love playing a sport or painting. They love their bed. They want rest and a moment without pressure. If this is you, save the exercises for later, and start with some self-care. Prioritize rest. Take your foot off the gas pedal for a few weeks at work. Have a bath, read a novel, stare out into space. Make a yummy meal. Give yourself permission to step off the treadmill of your life for a little, and see how you feel.

You most likely will find that the world doesn't fall apart. The DMN is the one putting intense pressure on you to live up to a certain standard and/or be productive every minute of the day. Separating yourself from that pressure separates you from the DMN, showing you that maybe the DMN doesn't have it all figured out. Maybe you don't have to grind endlessly at work, putting in fourteen hours a day, to be good at your job. Maybe you don't have to go to that party to be liked and accepted by your friends. Your kids will live if they don't have nutritionally balanced home-cooked meals every night. Whatever standards you give yourself, reconsider them. If you insist "I have to do it this way," it could be time to reevaluate.

I also see people give up if they don't have die-hard passions in their life. Sure, it's great when someone loves lacrosse, played lacrosse in their youth, and now coaches their kid's lacrosse team. That's clear. That is like having a favorite food and saying you could eat it for the rest of your life. But what if you didn't really have any great loves when you were younger? Or now? What if you feel that you've coasted, trying out everything and not minding anything, but not loving anything either? What do you do then?

For many, magic in life comes not from a singular pursuit, but from the fact that we somehow got born into these physical bodies on a planet spinning around a fireball in the middle of space that probably doesn't end and WTF does that mean that it doesn't end? Life itself is absolutely mind-boggling. Cells innately know how to divide and then become organs and then all work together so that you can look around and move where you want to and do stuff? It's crazy. And that can be a passion right there. So if you are someone who can't point to specific things that you love, I want you instead to notice what you notice. You read that right. What catches your attention? For you, it could be the peace of sitting in your car. The joke your colleague said. The taste of that chocolate. It could be simpler. And the simplest parts of life are the most magical. Continuing our "What's your favorite food?" analogy, these are the people who don't have a favorite food but appreciate tasting and experiencing all types of food.

Then there are the tapas people, who like small dishes. They become interested in something, do it for a while, then move on to something else. They often criticize themselves as being flighty or wishy-washy. Now we all know that's the DMN coming up with derogatory labels for no reason.

These individuals actually experience life as a sampling of yummy options. They try one, enjoy it for a while, then move on to the next treat.

When social media launched memes about how often men think about the Roman Empire, I found myself suddenly interested in the Roman Empire. I watched a bunch of archeology shows, marveled over the perfectly preserved leather shoes and wooden doors, and went down some Roman life rabbit holes online; then I was done. It was short-lived, only two weeks or so, but it was super fun while I was engaged in it. I let my true self lead me down a rather riveting path and back out again. I'm curious about what I'll be curious about next.

You may flit between these types of loves throughout your life. You may have times when you have one intense, burning passion. You may have others when you move into and out of passions rather quickly. And sometimes you may not feel passion for anything whatsoever. It could be because the DMN has taken charge of your chariot (okay, maybe I'm not totally out of my Roman Empire phase yet) and made you lose connection with your heart. Notice this, practice some self-care, and revisit your love list. If it is not the DMN, then those are the times when your heart is speaking more subtly. Maybe you notice the beauty of someone's shiny hair or the way the sun hits the grass. These expressions of the heart are just as important as the others.

Step Two: Notice the Nudges

I find it ironic and telling that for thousands of years humans have assumed that their inner monologues were the seat of themselves, while at the same time telling each other to trust their gut when it comes to important decisions. They listened to the voice in their head and then listened when that voice spoke from their belly. Talk about a DMN-dominated world.

When it comes to life decisions, big or small, stop listening to your gut brain, roiling over whatever your DMN is on about. Instead, listen to your intuition.

> **EXERCISE:** Listening to My Heart
>
> To connect with your heart's voice, start with some small-stakes matters; what to eat, which route to take to work, or what show to watch. Any small everyday decision you make can be passed through the heart test. Let's take the example of what to wear.
>
> When you go to your closet in the morning, I want you to drop into your heart. What is it telling you? Does it want you to wear a comfy sweatshirt or a pretty blouse? Which clothing gives you a spark of warmth and goodness? If you find that you are *thinking* (say, *I should wear this shirt because it hides my belly*), take a moment to stop your thoughts. You need to get the DMN out of the picture. Then reach back into your heart and feel for any subtle urges.
>
> If you hear crickets, try a different approach. Hold up two options. You could wear this shirt or this sweater. Feel into each, and see which one you're drawn to. If nothing comes up, then pick one. Say to yourself, *All right, you're wearing this.* If you feel a sensation of *Noooo! Wait!* that's your heart telling you what it wants. Hold up the other option, and pay attention. Notice your body, breath, and so on. That is your heart talking.

Wearing clothes is just one facet of self-expression. However, when we stop dressing as we think we *should* and instead dress how we really *want* to, interesting things happen. I knew one woman who went from wearing all black to big bursts of color. Another stopped wearing tight, fitted pants and chose flowy, comfy options instead. Another put away all her fancy high-end jewelry and started wearing crystal bracelets. This is one of the myriad aspects of life over which we do have a choice, and it is up to us whether to exercise it. Do we dress according to an ideal, standard, or trend, or what feels good to us personally?

When the heart says yes, you will feel light, airy, and energetic. Your body will feel calm. You may feel excitement or a sense of play. These are all signs that your heart is loving something. If you feel any sort of unease or discomfort, that is the heart telling you no. You may experience unpleasant emotions, negative self-talk, or physical tightness. The cardinal rule to follow is if it feels good, go toward it. If it doesn't feel good, don't.

Such a nudge happened to me recently. My OB-GYN said I needed to eat more plant-based proteins. She said, "Make a kale salad and put lentils on it." *Ugh*, I thought. *I hate lentils.* I smiled and nodded, wanting to be a good, attentive patient, but in my heart, I felt humming discord. *Lentils are not right for you*, it whispered.

Is this a story about how I listened to my heart? Heck no. I went right to the grocery store and bought lentils and kale. The kale sat in my fridge for thirteen days before I threw it out, and the lentils are still in my cupboard, unopened. In the days after that appointment, I ate lots of unhealthy things and beat myself up over having no self-control. Other women are eating endless beans and salads; what is wrong with me? I had stopped listening to my heart brain, and fell apart. I gave my DMN an opening to take the mic and tell me I lacked willpower and the ability to make sensible choices. Next time? I'm going with my heart brain, no matter how educated or convincing the advice sounds. It chooses eggs, or butter-lettuce salads, or other good things…that just aren't lentils. That is far healthier for me than following random health-related guidance.

Step Three: Pick Your Priorities

A mom one street over hosts wonderful parties. She gets the neighborhood together for block parties, first-day-of-school send-offs, and more. Whenever I receive one of her invites, I think, *Ugh. I should be hosting parties. I should be repaying people for inviting me to things. I should be putting myself out there. It's the right thing to do.* I feel tired, cranky, and put out, like throwing a party is some obligation I must drag my way through, complaining the whole time.

One time I saw her and said, "You throw the best parties." I said it because she does, and also to release some of the guilt and pressure that had built up in me from *not* throwing parties. She said, "Omigosh, I love party planning. It's so much fun. And I love getting everyone together. I haven't had one in a while. I should start planning my next one. Thanks for reminding me!"

I was floored. You *like* putting on parties?! It's *fun* for you? It was such a wake-up call.

We tend to watch what everyone else is doing and feel that we should be doing the same. Hello, DMN. We need to realize that different things light different people up, and that's what makes the world go round. Party planning should be a priority for my neighbor—but not for me.

EXERCISE: Prioritize!

To identify your priorities, connect with what creates the most meaning and love in your life. What do you authentically care about? What makes your heart center hum? What makes you feel good? Don't put thought into it—drop into your heart, and in your journal or notebook, list anything and everything that comes to mind. If you see repeats from the first exercise in this chapter, great. You are homing in on the things that make you *you*.

Now go back through your list and rank your entries. What is the *top* item that brings meaning, joy, and love in your life? That's number one. What's the second? Feel free to stop when you get to ten.

Take a look at your list. The top three items are the ones you want to incorporate and prioritize as much as you can. The next seven are the ones you build in when it makes sense, and the rest are gravy. If you can make them happen, great, but if not, that's fine too.

Once you have your priority list, make it as concrete as you can. As a personal example, I started with "My son." I love him more than anything, and he brings so much meaning to my life. When I made it concrete, though, it was *time with my child*. For this divorced mom, the knowledge that he spends every other weekend away from me sits heavy. Everyone talks about how quickly these years go by, and then I miss out on half of it? It's hard. So the priority I live by is to really be there when he's with me. If my book club schedules a meeting for when I have him, I decline. If I could get another email out, I choose not to. It draws a clear line in the sand that I can personally use to decide my actions, and a reason I can give others for turning down invitations.

Women's social conditioning to be amenable and please others can make it really hard to say no. Making a priority list creates clear boundaries so that you aren't being pushed and pulled in so many directions. It offers a shortcut, a decision-making framework so you don't have to drop into your heart every time you face a choice. Identifying your values and then living by them can make you feel strong. You aren't living your life according to other people's expectations or beliefs—you are staying true to yourself.

A priority for many? Sleep. Sleep makes us happier, more energetic and level-headed. Some say, "But sleep doesn't bring meaning to life! That can't be a priority!" They're wrong. If you love how you feel and how your life feels when you are rested, it sure as heck can be. People are more present when they are rested, and being present can lead to meaningful moments.

Some people do this exercise and realize that they hate small talk and want to put themselves in only those social situations where they can show up and talk freely and authentically. They prioritize fewer, more genuine friendships over bigger social engagements. Others recognize that they feel better when they exercise, and decide to schedule barre three times a week, arranging their work and family calendars around it. Or they realize how important their friends from growing up are, and that they need to make an effort to do an annual weekend away with them instead of letting time slip by. Others see that their favorite, most meaningful times were when they were having fun, yet they haven't laughed in a while. They decide to go to more stand-up comedy shows and read biographies of funny people.

Sometimes this exercise can lead to bigger life changes. Some find that the hangovers from alcohol are interfering with other priorities and decide to cut down. Others consider moving closer to their biological or chosen families. Still others change their lines of work to connect more with their true passions. There is a ripple effect that can happen when we identify our priorities and align our lives according to them.

We are all unique beings, with our own recipe for what lights us up. What is yours? Bring those elements into your life in every big and small way you can so your day-to-day experience is truly, authentically yours.

The Truth About Beliefs

As you are doing this exercise, be mindful of any beliefs that crop up and stand in your way. For example, you may love being in a committed relationship and want to make finding one a priority, but you hear your DMN saying, "You are more likely to get struck by lightning than to get married after forty." Or you love teaching kids, but you hear, "Money doesn't grow on trees." Or you love getting your nails done, but you hear, "Spending money on yourself is wasteful." The vast majority of beliefs are ossified DMN thoughts that you've heard from others or told yourself so many times that you now take them as gospel. These tropes are how the DMN runs the show. So when you come across one, recognize it. See how it pushes the clan agenda of being accepted, playing small, or staying safe. See it for the malarkey it is. If needed, state a new belief, preferably out loud: "There are many wonderful potential partners out there." "I have everything I need." "I love myself and the experiences I give myself." Take the time to make sure your hardened DMN thoughts don't stand in the way of you and your priorities.

Step Four: Go with the Flow

This chapter is all about living from a place of love, and it wouldn't be complete without looking at how to love ourselves. How do we practice self-love? On the podcast *We Can Do Hard Things*, Abby Wambach, one of the greatest women's professional soccer players of all time, gives a raw and authentic account of how hard it is to practice self-love. "My god," she says, "that should be something that is just innately in me, and I feel like such a failure. I know I'm a good person, but how could I not love myself? Why is that so fucking hard for me? It's not fair." By the end of the podcast, she is no closer to an answer: "But how will I find out for myself once and for all? What will be the thing? What is the way? I'll do anything!" I hear you, Abby. You are courageous to be so honest, and you are so not alone. We all want self-love, yet many of us don't have a good concept of what it looks, feels, tastes, or smells like.

Let's try a thought experiment. Think about owning a pet. I have a Shih-Tzu named Henry, and although he is ridiculously cute, it has not been a smooth ride. When he was a puppy, he pissed on every piece of formal living room furniture I own. When he's going to vomit, he beelines for a carpet and disposes of his fluorescent yellow, bubbly spit-up there. (Why not on the tile? Or wood? Why do it on the one thing that stains!?) He needs to go for walks on rainy days, cold days, sick days, and sad days. It is a lot. Yet I adore him. I accept him as he is, in all his doggy glory.

Self-love is accepting yourself, in all your human glory. You will mess up. You will get derailed. You will scream and shout and cry and get angry. You will feel sad and hurt, and you'll struggle to get through some days. You will hurt others, and you will be there for others. You will have boring times, scared times, and fed-up times. All of this is part of being human, and self-love is being kind to yourself through all of it. It's not just feeling proud of how you look when you do your hair and makeup and put on some cute clothes. Self-love is more like a deep underground current that flows no matter what the weather on the surface is like.

This can be really hard to do. We can be so quick to beat ourselves up for snapping at our families, having a messy house, or not getting as much done as we would have liked. We can feel shame when we get impatient with other drivers, or say something that didn't land well with a friend. We all have expectations for ourselves, and our DMN's instinct is to beat ourselves up when we don't achieve them. Even when we've got all the science, we can still feel disappointed and wish we had done better. We can still act unloving toward ourselves.

I believe that some of us—particularly women—have a hard time stepping into the experience of self-love and doing things in alignment with self-love because we are so focused on giving outward rather than inward. When we work, we want to give to our teams and projects. When we are at home, we want to give to our loved ones. Sometimes we give because it is what is truly in our hearts, but other times it comes from an overactive DMN that feels we need to do these things to be liked and accepted. Regardless of the origin, it takes practice to turn that focus around.

How do you express love to others? How do you take care of them? Bake for them? Shower them with kisses and hugs? Turn that same energy back onto yourself. If you love feeding others, feed yourself with the same care. If you love giving hugs and kisses, treat yourself to a massage.

If a friend messes up and does something bad, how do you support them? Do you show up with compassion and understanding? That is the energy to give to yourself when you make a mistake. I know a woman who saves worms on rainy sidewalks, moving them to open soil, who also overworks and drives herself to exhaustion. She directs all her love outward instead of inward. As you honor, care for, and respect others—honor, care for, and respect yourself. You are a powerful, brilliant version of creation, just as much as they are. When you open your love channel toward yourself, you will feel that current underneath that always flows, no matter what is happening on the surface.

If you mess something up, accept the mistake you've made and let it go. You matter more than the mistake. To be clear, when I say acceptance, I do not mean forgiveness. You do not have to forgive yourself or forgive others for their bad behavior. You did the best you could with the information and context of that moment, and so did the others around you. You all learned from it. Let it go. This is similar to rejecting your DMN thoughts. You don't argue or solve them—you let them go.

Caring for ourselves is reconnecting with what we love and bringing more of it into our life. It is listening to that quiet little voice in our heart that tells us what is good for us and what isn't. It is aligning our life to what we find meaningful, and standing strong in that, even if it goes against cultural and familial expectations. It is honoring, respecting, and accepting ourselves just as we strive to honor, respect, and accept others. For us women, it is not "Do unto others as you would do unto yourself"; it is "Do unto yourself as you would do unto others."

In the next chapters, we will see how to love and care for ourselves across the many contexts of life. As we all know, not all of life is a choice. We didn't choose our family. We didn't choose what happened to us as a child. We don't choose our mental health diagnoses. Nor do we choose the world around us, with its social media and geopolitical climate. We have to work with what we've got in many parts of life. Let's explore ways to stay true to yourself in everyday life—no matter what gets thrown at you.

Chapter 8

Staying True to Yourself in Relationships

Relationships are a cornerstone of DMN activity. They are the reason the DMN took center stage in the first place! The DMN helped humans assimilate into tribal culture so everyone could work together as a cohesive group. Nowadays, the DMN overlays those survival instincts onto the normal ups and downs of your relationships.

For many of us overthinkers, relationships can feel challenging and painful because the DMN loves to make them that way. After we go to a party, it ruminates on every last thing we said or did. Before meeting up with people, it goes on about what we should wear, who we will talk to, and more. Solitude can seem like the ultimate solution because it is less activating for the DMN, and many of us call ourselves introverts because it is easier to be alone with a book than to listen to the DMN harangue us about our social lives.

Imagine how it would feel to relate to others without the automatic, negative self-talk of the DMN. It would be amazing! We would take the offhand comments people say less personally. We would think less about whether others are judging us or if we said the right thing. We would relax and enjoy ourselves more, without needing alcohol or another substance to quell those negative thoughts.

We need to break the DMN's hold over our social lives. First, we need to recognize when it has the floor and is driving our insecurity, anxiety, and lack of comfort. Then we need to stop it in its tracks, knowing that it was meant for social situations thousands of years ago—not today.

Let's take a look at some common relationship situations that sound the DMN survival alarm in our head, and how to cope with them when they happen.

Feeling Left Out

There is a wonderful mom in my son's class whom I just adore. Let's call her Katie. She is a caring, inclusive, fun person. One night at an event, a woman in the circle said that she and I know each other because we are in the same book club. I watched as Katie's face dropped and uneasiness took over her body. *The moms she knew from school were part of a book club, and she didn't even know it existed? And no one had invited her?* Ugh. I still feel a punch in my gut when I think about it. Katie was left out.

Sarah had her day wrecked when she looked on Instagram and saw that a person whom she considered a close friend had a big themed party and didn't invite her.

The DMN was not built for the twenty-first century. In our hunter-gatherer days, we lived in one clan our entire life. Getting kicked out of that clan *did* mean life or death. But nowadays, we encounter myriad clans: at work, in our neighborhoods, in our friend groups, in our friends' friend groups, in our activities, and more. With all of these clans, there are a zillion ways for the DMN to perceive rejection. This endless, and honestly impossible, requirement to fit yourself into every social dynamic produces a constant hum of anxiety. *Am I fitting in? Am I liked? Am I part of the group?*

What do I need to wear/say/do to be accepted? You can drive yourself nuts this way.

And it doesn't stop with you. If you have children, you can spiral about whether they fit into their peer groups. *Do my kids have friends? Do we need to have more playdates? Do we need to do after-school and weekend soccer so he doesn't get left behind?* Our DMNs have a full-time job keeping us all safely enmeshed in "clans," and all we have to show for it is a lot of overthinking and unnecessary stress.

When we feel left out, we get a pit in the stomach. We feel nervous, on edge, and self-conscious. *Maybe people don't like me as much as I thought? Am I normal enough? Am I pretty enough, interesting enough, or successful enough?* We feel that everyone else has life figured out, while we are improvising. This cascade of DMN thoughts and corresponding feelings can lead to sadness, insecurity, anxiety, and loneliness.

If you are feeling the pain of rejection, in whatever context or form, try some self-talk. First, recognize that this flood of thoughts and feelings is not helpful or real. We have outdated human hardware that is misinterpreting the situation. We feel we are being judged, when in fact most human social stuff is more happenstance than that. In the book club example, one woman got invited because a member wanted to reciprocate for hosting a party. Another person got invited because they always talk books, and a member was like, "You would probably really enjoy my book club." Another person got invited because they just happened to be standing in the circle of women when the idea of the book club came up. The reality is simpler and more ordinary than we imagine, and we need to coach our DMN to not take these occurrences so personally. It is not a judgment on us at all.

Another trick is to talk to your DMN and tell it who you choose as your clan. For me, my clan is my son, my parents, and a few close friends. That's it. I have deep, wonderful friendships, with people whom I genuinely love, yet I know that at the end of the day, their clans probably don't include me either. They all have busy, complicated lives navigating both normal stuff (doctor appointments, work deadlines, holiday preparations) and hard stuff (aging parents, children who need help with learning and development). If I don't hear from a good friend for a year, no worries. If a friend talks about getting together with people and I am standing there, no problem. I love them, but whether I am included in their lives is not my concern. If my DMN takes the floor and starts stressing me out, I tell it that I haven't invited them over recently either, so sit down.

Let's take a look at the ultimate rejection experience: romantic breakups and divorces. These are your DMN's personal nightmares. When you enter into a relationship, you form a mini-clan with the other person. You share living spaces, secrets, and friend groups. You create plans and think about the future. You support each other in good times and bad. When a breakup happens, you are no longer worried about getting kicked out of the clan—it's happened! The DMN sees this as a Code Red situation and will spout all sorts of nasty things at you. It will also make you feel disoriented. You were a member of a very exclusive clan, and suddenly you are out? The DMN can't grasp that well. The whole world feels strange and different as a result.

When surviving a breakup or divorce, remind your DMN that the relationship wasn't your only clan—you are a member of a bunch of other clans (family, friends, coworkers). If your DMN starts spouting nonsense (*you're worthless, no one will ever love you again, you will die alone*), call out that those ideas are DMN thoughts, and do your best to distract yourself, even if all you can muster is watching a movie. Lean on the people who love and support you, to give your DMN some data points that you *do* still have a clan and are accepted by others. Even though you don't want to burden people, you need to reach out and get support. The more you can stop your DMN from running your breakup, the more you can focus on healing your heart and reconnecting with your true self.

Managing Criticism

Think of all the ways we can feel criticized in modern-day life. Performance reviews at work. Feedback on a paper or presentation. Report cards. A person looking down at their phone when talking to you. Someone who stops talking with you so they can say hi to someone else. Someone asking why you don't have a boyfriend. A partner who hasn't wanted to be intimate for a while. The list is endless.

Criticism can activate all categories of DMN thought at the same time. It's a DMN maelstrom. One bad performance review can make you feel you're not smart enough (*failure*), are going to be fired (*fit in*), and will end up homeless (*fret*). It can stress you out and make it hard to handle anything.

Jill's husband criticized her constantly: She didn't attend a good enough college, she wasn't providing their child enough stimulation, she was a boring stay-at-home mom, and so on. The offhand digs began to take a serious toll on her. She felt ashamed of herself, unworthy of love, and unable to do anything right. I mean, here was the guy who'd said marrying her would make him the happiest man alive, and now he was trashing her? He was her clan, and yet according to him, she was never good enough. Her self-worth went into the dumpster, and she suffered tremendously. It was a DMN nightmare.

The DMN loves criticism because it perceives it as truth that polite society would never say. Typically, people don't share the negative thoughts about others that pop into their heads. However, if someone tells us we aren't attractive enough to get a romantic partner, or we are parenting badly, the DMN latches onto that piece of feedback as gospel, thinking that if it corrects whatever the person implied about us, we'll be more secure. We tend to dismiss the positive things people say about us—thinking that they are just being nice—and cling to the negatives, because the DMN perceives these as the action items that will lead us to the promised land of clan inclusion and safety.

Unfortunately, it is dead wrong. When people criticize us, it tends to come from one of two places. Sometimes criticism is the person's own DMN made vocal. If their DMN tells them that they have to look good to have a boyfriend, it will tell you the same thing. Jill's husband, who implied that she was worthless because she didn't have a paying job, lived by that same creed. He would say things like "I need intellectual stimulation to be fulfilled," when his true motivation came from his DMN believing he would be kicked out of the clan if he didn't have a high-powered enough job to fit in with his family and peers.

When these criticisms come from our parents, they may be a backhanded way of trying to protect us. If a parent's DMN is obsessed with financial stability and believes the path to safety is becoming a doctor or lawyer, they are going to harp on their children if they don't pursue those professions. Children who instead chart their own course may feel they are unaccepted and unloved by their parents. They are actually right—and wrong. The parent's DMN doesn't accept or love them. Frankly, it doesn't accept or love anyone. That's not the DMN's job. Parents can have deep love for their children, yet not express it because they are leading with their DMN and not their heart.

The other common form of criticism we see in the world is bullying. The bully's DMN has decided that to fit into the clan, it is going to criticize and reject others. If you are the one deciding who is in and who is out, then you can't be thrown out, right? It is a particular form of safety to be the one dictating the group's social norms.

We talk a lot about how bullies are insecure and take it out on others. What we really should say is that bullies are particularly DMN-dependent. They lead their lives based on how to fit in and be accepted. Therefore they are quick to call out others when they do not. The more visible they make the clan norms, the more relieved their DMNs are that they are doing enough to fit in. Or the bully may lash out at someone different because, in their heart of hearts, they would like to be living the same way yet fear it's unacceptable to do so.

One more source of criticism is emotionally immature people. Psychotherapist Lindsay Gibson found that many of her clients told stories of relatives behaving like children. When Gibson and the clients discussed their behavior in terms of immaturity, it greatly helped them. Gibson created a framework for identifying emotionally immature individuals. Typically, they seek shallow, superficial relationships, lack boundaries, and dismiss or reject any overtures for deeper emotional connection. They play the victim, blame others for their bad feelings, and reject others' perceptions of reality. They will neither take responsibility for their actions nor engage in constructive communication on how to better care for the relationship.

Consider Anne. On the surface, she appears to have a great life: a solid job, a nice house, and lavish annual vacations with her picture-perfect family. Yet she will tell you that nothing ever goes well in her life. No one appreciates her at work. Others take credit for her ideas, and when they get promoted, she feels it should have been her. Talking about her parents, she rolls her eyes, calling them selfish and tiring. She is frustrated with her spouse, feeling that he should be more active. She calls him a loser. There is something wrong with every meal she eats unless she cooked it. Even her vacations are a sore topic: too many people at the airport, mold in her five-star hotel room so she couldn't sleep, too many flies on the beach. Rather than look within for joy and contentment, she looks to the world to satisfy her needs, and the world always comes up short.

For these individuals, their DMN has changed its orientation. It focuses outward, perceiving that others' actions and qualities are responsible for their safety and well-being. Their DMN spins endless tales of how the world is letting them down and everyone else is to blame for how they feel.

The following table offers a side-by-side comparison of how the DMN sounds different in emotionally immature people.

Inward-oriented DMN	Outward-oriented DMN
I'm not good enough.	Everyone in my life hurts and disappoints me.
I feel like I keep messing up.	You're doing this wrong, and this wrong, and…
I'm not as smart or experienced as my colleagues.	My boss isn't smart. He only got to where he is because of good timing.
I could end up in a bad place if I lost my job.	She should watch out, or else I will tell the world what she's done and make her pay.

These people may appear highly self-centered; in reality, they are outward-focused. They struggle to take responsibility for their actions and how they make others feel because, to their DMN, they are not to blame for how they feel. They believe everyone *else* is to blame. They struggle with relationships because, to them, it is about what the relationship is doing for them rather than what they can do for the relationship. They tend to not seek treatment, because their DMN believes the world needs treatment, not them. It is as if their true self is absent. Perhaps their true self has been so damped down that they cannot connect with it.

When you are in close contact with an emotionally immature individual, it is like getting a one-two DMN punch. First your DMN is doing its normal thing of evaluating your every move. Then you have the other person's DMN doing the same thing, in a dismissive and cruel way. Your DMN then spins on the other person's criticisms, validating the nonsense your DMN said in the first place (*I am not worthy of love, I am not good enough*). If your emotional well-being was a weather pattern, it would be like two major storms meeting and turning into a superstorm.

How do you take care of yourself in the face of criticism and when dealing with bullying or others' emotionally immature behavior? How do you turn down the DMN noise, reconnect with your true self, and feel surer of yourself? Try these ideas:

- Keep in mind that criticism is often not reality or truth, but a reflection of the other person's DMN-driven insecurities and fears. Either the person is sharing something their DMN tells them all the time, or their DMN isn't functioning properly and is harping on you instead. Regardless, these comments are coming from an automatic, outdated survival machine. Just because they come from their DMN and not yours doesn't make them any more valid.

- Take a step back and assess the "truthiness" of the criticism by asking your heart whether the comment was valid. Since criticism is often so DMN-driven and fear-based, most of it won't hold up to this test. If someone made a snide comment about your hair, what does your heart say? Probably that it likes your new look. If a friend tells you that you hurt their feelings when you interrupted them, what does your heart say? Probably that it sucks to be interrupted and you should apologize. Listen to that little inner voice. It knows what is best for you. It may say the exact opposite of what your DMN says. Don't worry. In that debate, your heart always knows the best path forward.

- Soothe yourself with connection. Criticism makes us feel like our membership in the clan is in question. To combat this false perception, give yourself a clan experience. Tell a trusted friend what happened, and let them validate your feelings. When we feel safe and connected with others, it calms down our nervous system and assuages our fears.

- If you don't feel comfortable sharing what happened with someone else or there isn't an opportunity, provide that connection to yourself. This will calm the stress response, bring your CEN back online, and help you move past the experience:

i. Close your eyes, take several deep breaths, and still your body.

ii. Imagine your higher self in a rocking chair. You are big, bright, and powerful. You are endless love and acceptance.

iii. Picture your current self as a young person, with all the feelings you currently have. Your higher self asks what's wrong.

iv. Climb into your higher self's lap, and share your story.

v. Feel yourself being rocked and hugged, soothed and loved. Sit in the feeling for as long as you like, or at least until you feel your body is calmer and your breath has naturally resumed at a slower pace.

vi. Revisit this place anytime you start to feel insecure or anxious.

You can also achieve this same experience from a guided meditation on self-love or one that releases fear and anxiety. Choose one that resonates the most with you in the moment.

- Make a list of people from whom you will take criticism. I used to work at a firm where "feedback" was the gold standard. You got feedback every day, often multiple times a day. After overstressing about feedback and burning out, I decided to try a different approach. I made a list of colleagues whom I admired and trusted. They were great at their jobs and cared about the people with whom they worked. When they gave me feedback, I listened. I asked how I could do things better. I cared. When other, less-liked superiors gave me feedback, I smiled and nodded. I said, "Thank you for the feedback," but I actively tuned them out. My DMN often created a little storm based on their input, but then I told myself to let it go and remember what made me good at my job. It helps to keep a list of what makes you good at your job to go with the list of people from whom you will take criticism. This applies to all parts of your life: friendships, marriage, family dynamics, parenting, and so on.

- If you are dealing with an emotionally immature person, the solutions shift. Operating with a dysfunctional DMN, they won't hear reason or empathize with how they made you feel when you try to resolve a hurtful situation. Instead, they will turn the narrative around to make it appear that you are the aggressor. Since their DMN will not consider their responsibility in the situation, the best way to cope is to remove yourself from their line of fire. It's not a real relationship. You may have to expect less from the interactions or even let the relationship go entirely. For more information on how to manage these situations, check out Lindsay C. Gibson's book *Adult Children of Emotionally Immature Parents: How to Heal from Distant, Rejecting, or Self-Involved Parents.* She shares insights on how to take care of yourself while navigating these relationships.

Unless you are made of steel, rejection and criticism will make you on edge and unsure of yourself. These situations activate your DMN, making you doubt yourself and your choices. Knowing that these responses are automatic and were useful at some point in history can help you take a deep breath and let go of the insecure thoughts. Think about how you startle at an unexpected loud noise. You laugh afterward. The human body is funny. Keep this analogy in mind when you feel hurt from a social situation. Your mind just heard a figurative loud noise and is doing the verbal equivalent of a startle reflex. Try to step back from the thoughts and feelings and let them go.

Chapter 9

Scrolling and Trolling: Navigating Social Media

For better or worse, social media is now an integral part of life. Small business owners and creatives use social media to reach customers and audiences. Teenagers feel they need to be online because so much socializing happens there. Some use it to stay connected with family and friends far away. Others appreciate the funny content, the connection with admirable thought leaders, and the greater community that social media provides.

Unfortunately, there are negative ramifications to using social media. It can lead to low self-worth, depression, anxiety, and even suicidal thoughts. Since social media is here to stay, we need to understand what it is doing to our brains and what choices we can make to minimize the negative effect. When we overlay an understanding of the DMN on top of social media research, we gain a view into what is going wrong and how we can handle it better. In this chapter, we will look at two of the most destructive aspects of social media: scrolling and trolling.

Researchers distinguish between active use—messaging and engaging with people directly—and passive use, aka scrolling. They find that negative outcomes from social media tend to relate to passive, not active, use. If we think in terms of the DMN, this finding makes complete sense. When we are focused on what we are saying to a friend, the DMN is offline. Scrolling, on the other hand, is a DMN candy store. The DMN lives for this! It takes in all of the data from others' appearance, friend groups, lifestyle, professional expectations, and so on, and tells us that we should be doing and having the same.

As you look at the endless, heavily edited photos and videos, the DMN perceives social standards that no one could live up to. It tries to answer its cardinal questions—"Do I fit in?" "Am I good enough?" "Will I be okay?"—and finds you falling short on all measures. It then convinces you that you are a boring, drab, unsuccessful member of the human race. Everyone else seems to have happier, bigger families, more successful careers, more hobbies, more friends, better vacations, and better looks than you. Research shows that when you negatively compare yourself to others on social media and then listen to your DMN talking about it, you are more at risk of developing depressive symptoms. While journalists call it "Facebook depression," we can see that scrolling strengthens the DMN's pathways in the brain and weakens the CEN, leading to the mental health symptoms just listed.

The Social Media Scrub

How do you adjust your scrolling behavior to protect your brain and well-being? First, revamp your platforms. Either start all over again with brand new accounts or do a heavy edit of your existing ones. Remove any acquaintances who give you a queasy feeling when you see their content. If you get stressed and anxious when looking at life updates from the people with whom you grew up, unfollow them. If content from leaders in your industry makes you feel nervous about your own work situation, unfollow them. Be ruthless. Our DMN may tell us that we need to stay up-to-date on every last news item, but that's false. The world will keep spinning.

Next, follow people who make you feel good. They may be inspiring, thought-provoking, super-authentic, hilarious, or all of the above. Prioritize whatever content makes you feel good.

Notice how and when you scroll. Most people do it as a way to release a little pressure, give the brain a break, or kill time, as a matter of habit. Target the last two first. When you are using social media to kill time, come up with something else to do instead. I find mind games like Wordle and Scrabble are a great replacement, since they require focus and strengthen the CEN. If you have something else to reach for, it will be easier to avoid social media. As an alternative, think about what you will have for dinner or what's on your to-do list for tomorrow. If you appreciate the comfort of looking at a screen, write yourself an email with those dinner ideas or to-dos. Seeing those moments as opportunities for productive planning can make them seem like less of a waste. For the times that social media use is habitual, also find a replacement activity. Instead of scrolling before bed, read a book or watch television. Getting your brain out of the habit of endless, ever-changing, fast-paced stimuli is a helpful step to reducing the DMN's grip on your life. You can also make it harder to access by deleting the apps from your phone so you must log in from a browser. Then, when you do log in, you have a moment to check whether you really want to scroll.

When you seek the distraction of social media because work is super stressful or your DMN is going on about something annoying or you are just plain exhausted and want a few laughs, you can use it without feeling that you are overusing it. I like to think of social media as dessert. If you eat well most of the time, having an occasional chocolate cake or ice cream won't hurt. If you actively focus on other things 95 percent of the time, spending 5 percent of your time on social media can't do much harm.

Monitor yourself as you play around with what social media content you see and when. Some people may feel okay if they scroll for an hour a day, while others may feel like crying every time they go on a platform. If you have a negative reaction to social media, get off of it. It is not worth risking your mental health for the sake of sensationalist content. Keep up-to-date on what matters most to you through newspapers, magazines, books, and other paper-based material. Connect with your friends through texts. Set up the boundaries that work for you and your health.

The Problem with Trolls

Now let's turn to trolling. In her Netflix special, *The Call to Courage*, Brené Brown, a researcher who studies shame and vulnerability, shared the comments she received when her first TED talk went viral on YouTube: "Less research, more botox." "Of course she embraces imperfection. What choice would you have if you looked like her?" "She should wait and talk about worthiness when she loses 15 pounds," "I hope someone kills her," "I feel sorry for her husband and children," "She's what's wrong with the world today." She was overwhelmed by the aggressively negative response her talk received. To numb out the terrible DMN thoughts and feelings, she spent the day on the couch watching *Downton Abbey*. Go Brené! It was smart of her to shut down the DMN by escaping into a turn-of-the-century British drama. Negative feedback online is a gut punch to the DMN, triggering all of our worst fears about being shamed and shunned. Knowing how destructive it can be, the question stands: Why are people so mean online? Why do they say such hateful, violent, and offensive things?

Jaron Lanier is a technology industry elder. He founded virtual reality as we know it today, and in 2018 was named one of the Top 25 Most Influential People over the last twenty-five years of technology history by *Wired* magazine. In his book, *Ten Arguments for Deleting Your Social Media Accounts Right Now*, he takes a compelling and raw look at how even he developed troll-like tendencies online: "After a short while, I noticed that I'd write things I didn't even believe to get a rise out of readers. I wrote stuff that I knew people wanted to hear, or the opposite, because I knew it would be inflammatory. Oh my god! I was back in that same place, becoming an asshole because of something about this stupid technology! I quit—again." He theorizes that, "With nothing else to seek but attention, ordinary people tend to become assholes, because the biggest assholes get the most attention." I would put this differently. Social media is already a place where the DMN leads. It loves the experience of passively sitting back and collecting information on what we need to do to fit in and be good enough in the world. When we perceive derision as the clan norm—aka the secret ingredient to generating likes and followers—the DMN embraces it and pushes it as far as it will go. Jaron was writing "what people wanted to hear," not how he really felt. He was writing the automatic vitriol his DMN produced to appeal to the DMNs of others. Wayne Dyer notes that when you squeeze an orange, you get orange juice. Well, squeeze a DMN and you get cruelty, directed toward either yourself or others. This is the true nature of the DMN. Social media just shows us how cruel it can be.

Part of the problem is that the DMN was never meant to operate in a vacuum. Early humans did everything together—eating, working, resting. We had real-life social networks, providing the checks and balances needed to stay productive members of our communities. This is true at both ends of the spectrum: We rely on our friends to help us feel better when our DMN tells us we're a loser, or to tell us when we've crossed a line, behaved badly, and need to make amends. These checks are critical. Watching someone's facial expression as you say something lets you know whether it is appropriate or not, a learning you then carry forward into future conversations. Knowing we could lose a relationship forever if we behave too badly keeps our behavior in check. The online environment offers zero checks and balances, making it a potential DMN feeding frenzy of inappropriate comments, untrue statements, and hate speech.

When the DMN operates without the counterbalancing input from in-person social interactions, we can either become terribly destructive, as in the trolling behavior just discussed, or feel kicked out of the clan. We believe what the commenters say. We take their offensive remarks to mean that we are worthless and don't belong. Like a sick animal who goes into the underbrush to die, we even entertain thoughts about how much better the world would be without us. There are so many cases of child and teen suicide directly linked to social media that the phenomenon has a name: *bullycide*. It is a sad reality of our times that lives are actually lost due to interactions on social media.

How to Limit Troll Communications

To protect yourself from the damaging influence of trolls, turn off comments. Give people other ways to connect with you, via email or direct messaging. Or stop posting altogether. We don't all have to be public figures. If you must post and view the comments—as many creatives and small business owners do—then consider hiring someone to filter your social media for the important, productive messages. If that's not feasible and you must read comments, give yourself some strategies to deal with the slings and arrows. One media personality I know does participate in her comments section, and she has a quick shorthand for how she dismisses the vitriol. Any time she comes across something negative, she says to herself, *Sucks to be you.* This keeps her in the frame of mind that people who spout nastiness are the ones who need help—not her.

When a derogatory comment hits close to home, go back and reread the criticism section in chapter 8 and try some of the techniques there. People who are rude online are not your clan. Use your knowledge of the DMN and the brain to dismiss what they say.

Interacting with social media is an inevitability for many of us, and things can go very awry in the online space. Knowing how social media use feeds the DMN, we can shift the look and feel of our social media to better support our mental health, and monitor our usage so it doesn't bring us down.

Chapter 10

Healing in Therapy and Beyond

By this point in the book, you see that the voice in your head is not you. You are not your insecurities, worries, or self-judgments. For an added healing bonus, consider working through these thoughts with a mental health professional. Therapy can also help when we have more to heal than just thoughts.

For example, Abigail loved the idea of separating from her inner voice, but as she did, she found one line of thinking that she couldn't shake. "I feel like I have my mother sitting on my shoulder. Whenever I feel like crying, I hear her say, 'Suck it up. Other people have it worse.' Or if I want to do something nice for myself, like get my nails done, I hear her say, 'What a waste of money.' I can dismiss my own thoughts, but I am struggling to dismiss her. I feel like I am rejecting her and her love if I do that." She decided to seek out an objective, caring therapist to process her relationship with her mother.

Perhaps you are struggling in your marital relationship. Or feeling disconnected from yourself and who you are. Or experiencing the symptoms of a mental health condition. Therapy is a place to establish a secure, trusting relationship with someone. When the going gets rough, having a "clan" in a psychologist, social worker, or therapy group can give you the strength and support to heal and move forward.

In this chapter, we will explore how the major branches of modern therapy all support clients in releasing DMN thoughts and/or weakening the DMN. Then we'll explore what trauma does to the brain and body (and the DMN), and the therapy approaches specifically designed to address those needs. With knowledge about how therapy works and what it can offer, you can make an informed decision on whether it is a good fit for you.

Cognitive behavioral therapy (CBT) involves identifying negative thoughts and behaviors, analyzing what caused them, and questioning their validity. Then the client and therapist work together to generate new, more adaptive thoughts and behaviors to take their stead. This treatment option could appeal to you if you want a sounding board to identify the triggers that cause your DMN to spiral, and to test the validity of what your DMN tells you.

Internal Family Systems (IFS) therapy says that each of us is a system made up of parts. The *inner child* is who we were when we were little, *managers* set up rules and habits to keep us running effectively in our environments, and *firefighters* protect us when we are in real or perceived danger. Separate from these parts is the Self—the nonjudgmental, observing part that is all-loving and all-knowing. As a client connects with the Self, they can process their thoughts and feelings with love and understanding. They learn to appreciate how hard their other parts were working to keep them safe. The manager and firefighter parts can back off a bit, knowing that the Self is in control. I see great value in this approach because it helps people separate from their DMN-driven thought patterns and reconnect with their true selves.

Take Brenda, who gives her all at work and pushes herself to the brink of exhaustion. Her manager part believes she must work super hard to survive. Put differently, her DMN spins tons of fretting thoughts about how she'll be doomed if she doesn't work hard enough. IFS therapy helped Brenda separate from her DMN/manager cognitions and connect with her true self in a lasting, meaningful way.

Another treatment option that is growing in popularity is psychedelics. Leading psychedelic researcher Robin Carhart-Harris has shown that psychedelics reduce activity in the DMN, which can have profound effects on individuals' mental health symptoms and general well-being. In a small feasibility study, twelve people received psilocybin treatment for persistent depression. These individuals had tried two or more depression treatments with no success. The results were startling. Everyone showed improvements in their symptoms. Two-thirds of participants were depression-free after a week, and seven out of twelve still showed improvement in their symptoms after three months. Carhart-Harris theorizes that treatment-resistant depression stems from what he calls "rigidity" in the neural pathways. When we get stuck thinking the same thoughts over and over, with the same response patterns, we slide into mental health states like depression and anxiety. Psychedelics break down those well-worn pathways by creating a *hyperplastic* state in the brain where rapid, profound changes can occur, creating new pathways and ways of being. This study was small and there is much more research to do before we have a concrete view on psychedelic efficacy, but it is promising.

Even hypnosis works by shutting down the DMN. Brain imaging research shows that when a hypnosis therapist says, "Now I want you to focus on my voice," it activates the client's CEN, and their DMN shuts down. Clinicians purport that hypnosis creates a safe state in which to process hurts and build new habits. This makes sense, since the DMN— our fear-driven, survival-centered network—is turned off.

When my DMN was on overdrive during the divorce, I went on an antidepressant and experienced great relief. Research shows that the DMN is predominantly serotonin-modulated, and SSRI medications reduce connectivity throughout the DMN. For me, it felt like it quieted my DMN. It didn't run as hard or say such mean things. I felt less overwhelmed. I slept better and felt less stressed. It somehow got me out of my downward DMN spiral, and I could begin to rebuild my life. Then, once I discovered the research covered in this book, I tapered off of my medication. I realized that I could dismiss my now weaker, less dominant DMN, and I didn't need the medication to do it for me. I had the tools to ignore it myself.

Many people feel an aversion to seeking out therapy, or taking psych medications, fearing that it means they are unhinged. Try reframing these options as a personal trainer for the brain. When the DMN is super strong, therapy or medication can be an effective way to dampen its functioning, helping you access your CEN and your true self. This is a brain thing, a networking thing—not a sanity thing, or some mark on your character. Some of us may require that extra scaffolding for the rest of our lives, and that is okay too. Every person's brain and nervous system are different and need different things to achieve well-being. If you have pursued these options, in the short or long term, pat yourself on the back for making a choice that supports you, your brain, and your quality of life.

The Challenge of Trauma

Now let's discuss trauma, which has its own unique profile and treatments. Leading trauma expert Bessel van der Kolk explains that trauma occurs when one is unable to escape a dangerous situation. While many think this means surviving a natural disaster or car accident, there are other, more subtle forms of trauma, like being parented by an emotionally distant parent. It is the lack of control and inability to change the situation that pushes the experience into trauma. When a person can take action, moving or doing something to protect themselves, they tend to recover more quickly and easily than when they cannot.

Traumatic events restructure the brain and nervous system. During a traumatic episode, in an immense feat of self-preservation, the brain shuts down most functioning. The CEN and DMN go offline. The hippocampus, responsible for processing and storing memories, also goes dark. It reminds me of when your computer glitches and asks if you want to start it in "safe mode." When people are in a trauma-induced "safe mode," they feel removed from the situation. Some even report floating above their body. All of this serves to protect the person from fully experiencing the pain and suffering of the moment.

Unfortunately, there are long-term consequences. Memories of the event are not stored normally, but are distributed across the brain and body as a collection of sensory fragments. Later, a flash of an image, a sound, a smell, or a physical sensation that mimics the traumatic experience can activate a memory fragment, which then activates the other fragments. Any cue that activates one of the fragments floods the survivor with all of the fragments, making them relive the experience as if it is happening in the present. *Flashbacks* are a hallmark of post-traumatic stress disorder (PTSD).

The second byproduct of trauma is a resetting of the stress system. Stephen Porges, the father of polyvagal theory, was studying heart rate patterns when he somewhat accidentally discovered that the stress system wasn't quite as simple as we first imagined. In the old view, our stress system consisted of the fight-or-flight response. Porges found that there are actually three nerve pathways in the stress response system, and they operate in a hierarchy. The first pathway creates a freeze response—the feeling that the world is just too much and we want to go to bed. Think of a mouse that stops moving when it hears a hawk, or how goats fall over when they are scared. Next is the fight-or-flight response, when the system perceives that we can survive if we get the hell out of there or land a killer punch. The last, most sophisticated pathway is active in times of safety. Called the *social engagement system*, this pathway exists only in mammalian species. We feel at ease and connected with others, and we engage in prosocial, community-based behavior. We feel like getting together with family and friends. Talking. Laughing. Sharing food.

Normally, we move in and out of these stages rather seamlessly. Get cut off in traffic? A fight-or-flight response ensues. Your heart races, and you may scream a profanity or two. Then you refocus on the podcast you were listening to, and your system settles down. Or you have a bad day at work, head right to the couch to binge-watch television, and feel better the next day. We flex in and out of these states, and hopefully spend a good deal of time in the social engagement state, feeling safe in our body and world.

Trauma disrupts and intensifies these states. When you get cut off in traffic, you can't recover as quickly. Normal things that didn't bother you in the past—such as driving over bridges or watching your kids climb a particularly high playground structure—now make you panic. You get stuck on one of the lower rungs of the stress ladder and feel immobilized, emotional, or jumpy.

You didn't choose any of these feelings or reactions. You didn't choose the hard things that have happened in your life. However, you can choose to work on them and heal from them. This is where treatment can help.

First, trauma therapy can address the body. Deb Dana, a therapist familiar with polyvagal research, created specific protocols to teach clients how to identify and move between the freeze, fight-or-flight, and social engagement systems. Another approach is *somatic experiencing*, a body-oriented therapy that helps the client embody their physical self in a secure environment. During the original trauma, the brain's "safe mode" may have disconnected them from their body's sensations. This therapy brings the client back into their body in a protected way. It starts with building awareness of physical sensations, such as pain, muscle tightness, or shortness of breath. Then the client works on calming the body and shifting it in and out of stress states, as one would stretch a muscle to make it more flexible.

Trauma therapy can also address the mind. Eye movement desensitization and reprocessing (EMDR) is our current gold-standard approach, helping people to integrate the sensory fragments of trauma into a cohesive whole so they can be stored as a memory. First, the therapist and client discuss the client's history, big events, and goals for therapy. Then together they identify several experiences they would like to process. For each, the client brings back the feelings and sensations of the trauma (only as they are comfortable—sometimes clients choose to start with less intense memories to get used to the process). Then, while the client is in that heightened state, the therapist moves their finger back and forth, asking the client to track it with their eyes. It sounds deceptively simple, but shifting your eyeballs back and forth provides bilateral stimulation to the brain that calms and soothes it, enabling memory processing and storage. Francine Shapiro, the founder of EMDR, hypothesized that EMDR works because this motion mimics REM sleep, a time when we also consolidate and store memories. As the prefrontal cortex reengages, clients find that they experience rational thoughts about their role in the situation (*It is not my fault that Dad left my mom*) and the roles of others (*With the way Dad was raised, he was doing the best he could*). When the DMN comes back online, the experience can finally be coded in the brain as *past*, making flashbacks less likely to occur in the future. Clients note how EMDR takes the emotional charge out of distressing memories and helps them move on with their lives. It has done wonders for individuals with PTSD. For more information on what trauma does to the brain, and a comprehensive account of treatment options, I recommend Bessel van der Kolk's book, *The Body Keeps the Score*.

If you seek out therapy or currently use any of these therapy options, I am proud of you. Don't downplay this: It takes tremendous bravery and strength to take that step and to work through those memories, relationships, experiences, feelings, and more. I hope you'll use this book and the exercises therein as a tool to support you in your process and journey.

Conclusion

The everyday experience of tuning out the DMN is subtle. You think a self-berating thought, and then a little shot of awareness goes *Nope. Not today, DMN.* Or you are sobbing your heart out while your mind spews nasty things, but deep down, you know they're not true. Gaining this knowledge is like installing a little referee in your head who throws a yellow card every time your mind turns unkind.

Over time, these small changes can have a profound impact. Imagine all the little moments every day and night that your DMN says something berating or belabors a situation. When you dismiss the noise, or at least know the noise isn't accurate, who you are isn't getting chipped away at anymore. Your self-worth isn't getting destroyed. The DMN's proposed solutions—to shape-shift to become something you're not, or to work so hard that you burn yourself out—lose their appeal. As you listen to your heart more and focus on the parts of life that bring you joy, meaning, and love, you build your sense of self and self-worth. It is a beautiful thing.

The way I see it, humans have exceptionally complex, sophisticated hardware. We are like souped-up sports cars, with lots of systems and controls. We feel so impressed by these systems that we let them take control. When the car flashes a warning sign, we fill it with gas. When it tells us to turn, we turn. In the midst of all this information and signaling, we forget that we are the ones driving the car. We don't have to do whatever the car tells us to do! We can use it to get where we want to go. And that is how we should treat our brains. Instead of looking to the DMN to tell us who we are and what we should do, we should use our brains to pursue what matters most to us. I get how we got it wrong. Now it's time for us to get it right.

Look around you: You may be amazed by how much DMN chatter you hear. Not just in your own head, but from world leaders, the media, online, your friends, your children. It is everywhere, and it is responsible for not only our individual pain and struggle, but our global pain and struggle as well. We have let this automatic, negative, fear-based voice direct our lives, our institutions, our politics, and more—because we didn't know any better. The DMN has tried so hard to keep us prehistorically "safe," yet it has contributed to most, if not all, of the troubling issues we face today.

Ironically, the voice that is dividing us, putting us and others down in our heads, is shared by all. Overcoming our most destructive tendencies is not an "I" game, but a "we" game. It is time for all of us to identify this voice for what it is—and dismiss it. To move out of the place of fear, lack, unworthiness, and self-recrimination, and show up as we truly are. When we accept and love ourselves and each other, we stop creating artificial out-groups. We can use our brilliant CENs to focus on what we care about most and come up with real change, for ourselves, our societal structures, and our planet. We can do this. Without our DMNs running the show, we are unstoppable.

Acknowledgments

Working on this project has been a dream come true, and I owe it all to Jed Bickman. Thank you, Jed, for believing in me. Thank you for your guidance every step of the way. I feel so grateful to have had your keen mind and caring heart steering this ship.

Callie Brown and Kristi Hein, your edits improved the book immeasurably. Hope Payson, your comments were instrumental. Thank you for bringing your profound expertise to this project! To the early readers—Joanna Barsh, Susan Sullivan, Caroline Fleck, Hara Estroff Marano, and Kaytee Gillis—I appreciate your taking the time to review the work and offer your thoughts.

Buzzy, I am so grateful to be your mom. You are an incredible person, and it is great fun watching you navigate your world with passion, kindness, and honesty. I love you to the end of space and infinity times back.

None of this would have been possible without my mom and dad. You have given me steadfast support and love, and it means everything. Thank you for being there through it all.

And to you, dear reader! If you picked up this book, we have something deeply in common and could probably talk for hours. Thank you for taking this journey with me.

Bibliography

Al-Shargie, F., R. Katmah, U. Tariq, F. Babiloni, F. Al-Mughairbi, and H. Al-Nashash. 2022. "Stress Management Using fNIRS and Binaural Beats Stimulation." *Biomedical Optics Express* 13(6), 3552–3575.

Armour, J. A. 2004. "Cardiac Neuronal Hierarchy in Health and Disease." *American Journal of Physiology-Regulatory, Integrative and Comparative Physiology* 287(2), R262–R271.

Aronson, J., M. J. Lustina, C. Good, K. Keogh, C. M. Steele, and J. Brown. 1999. "When White Men Can't Do Math: Necessary and Sufficient Factors in Stereotype Threat." *Journal of Experimental Social Psychology* 35(1), 29–46.

Bangasser, D. A., and B. Wicks. 2017. "Sex-Specific Mechanisms for Responding to Stress." *Journal of Neuroscience Research* 95(1-2), 75–82.

Bauer, C. C. C., S. Whitfield-Gabrieli, J. L. Diaz, E. H. Pasaye, and F. A. Barrios. 2019. "From State-to-Trait Meditation: Reconfiguration of Central Executive and Default Mode Networks." *eNeuro* 6(6), 1–17.

Baumeister, R. F., and A. Steinhilber. 1984. "Paradoxical Effects of Supportive Audiences on Performance Under Pressure: The Home Field Disadvantage in Sports Championships." *Journal of Personality and Social Psychology* 47(1), 85–93.

Beauchene, C., N. Abaid, R. Moran, R. A. Diana, and A. Leonessa. 2017. "The Effect of Binaural Beats on Verbal Working Memory and Cortical Connectivity." *Journal of Neural Engineering* 14(2), 026014.

Beilock, S. L., and T. H. Care. 2005. "When High-Powered People Fail: Working Memory and 'Choking Under Pressure' in Math." *Psychological Science* 16(2), 101–105.

Berman, M. G., D. E. Nee, M. Casement, H. S. Kim, P. Deldin, E. Kross, et al. 2011. "Neural and Behavioral Effects of Interference Resolution in Depression and Rumination." *Cognitive, Affective, & Behavioral Neuroscience* 11(1), 85–96.

Bratman, G. N., J. P. Hamilton, K. S. Hahn, G. C. Daily, and J. J. Gross. 2015. "Nature Experience Reduces Rumination and Subgenual Prefrontal Cortex Activation." *Proceedings of the National Academy of Sciences of the United States of America* 112(28), 8567–8572.

Brewer, J. A., P. D. Worhunsky, J. R. Gray, Y. Tang, J. Weber, and H. Kober. 2011. "Meditation Experience Is Associated with Differences in Default Mode Network Activity and Connectivity." *PNAS* 108(50), 20254–20259.

Burnell, K., M. J. George, J. W. Vollet, S. E. Ehrenreich, and M. K. Underwood. 2019. "Passive Social Networking Site Use and Well-being: The Mediating Roles of Social Comparison and the Fear of Missing Out." *Cyberpsychology: Journal of Psychosocial Research on Cyberspace* 13(3).

Capobianco, L., J. A. Morris, and A. Wells. 2018. "Worry and Rumination: Do They Prolong Physiological and Affective Recovery from Stress?" *Anxiety, Stress, & Coping* 31(3), 291–303.

Capuron, L., G. Pagnoni, M. Demetrashvili, B. J. Woolwine, C. B. Nemeroff, G. S. Berns, et al. 2005. "Anterior Cingulate Activation and Error Processing During Interferon-Alpha Treatment." *Biological Psychiatry* 58(3), 190–196.

Carhart-Harris, R. L., and K. J. Friston. 2019. "REBUS and the Anarchic Brain: Toward a Unified Model of the Brain Action of Psychedelics." *Pharmacological Reviews* 71(3), 316–344.

Carhart-Harris, R. L., M. Bolstridge, J. Rucker, C. M. Day, D. Erritzoe, M. Kaelen, et al. 2016. "Psilocybin with Psychological Support for Treatment-Resistant Depression: An Open-Label Feasibility Study." *The Lancet Psychiatry* 3(7), 619–627.

Carhart-Harris, R. L., D. Erritzoe, T. Williams, J. M. Stone, L. J. Reed, A. Colasanti, et al. 2012. "Neural Correlates of the Psychedelic State as Determined by fMRI Studies with Psilocybin." *Proceedings of the National Academy of Sciences of the United States of America* 109(6), 2138–2143.

Centers for Disease Control and Prevention. 2021. *Youth Risk Behavior Survey.* Available at http://www.cdc.gov/YRBSS.

Chee, M. W. L., J. C. Tan, H. Zheng, S. Parimal, D. H. Weissman, V. Zagorodnov, et al. 2008. "Lapsing During Sleep Deprivation Is Associated with Distributed Changes in Brain Activation." *The Journal of Neuroscience* 28(1), 5519–5528.

Colzato, L. S., H. Barone, R. Sellaro, and B. Hommel. 2017. "More Attentional Focusing Through Binaural Beats: Evidence from the Global-Local Task." *Psychological Research* 81(1), 271–277.

Dana, D. 2018. *The Polyvagal Theory in Therapy: Engaging the Rhythm of Regulation.* W. W. Norton & Company.

Dong, X., H. Qin, T. Wu, H. Hu, K. Liao, F. Cheng, et al. 2018. "Rest But Busy: Aberrant Resting-State Functional Connectivity of Triple Network Model in Insomnia." *Brain and Behavior* 8(2), e00876.

Doyle, G., A. Wambach, and A. Doyle. 2023, June 1. "The Bravest Conversation We've Had": Andrea Gibson [Audio podcast episode 215]. In *We Can Do Hard Things.*

Doyle, G., A. Wambach, and A. Doyle. 2023, November 14. "Abby Asks, 'Why Can't I Love Myself?'" [Audio podcast episode 258]. In *We Can Do Hard Things.*

Dunbar, R. I. M. 1993. "Coevolution of Neocortical Size, Group Size and Language in Humans." *Behavioral and Brain Sciences* 16(4), 681–735.

Dunbar, R. I. M., and S. Schultz. 2007. "Evolution in the Social Brain." *Science* 317(5843), 1344–1347.

Dunn, B. D., H. C. Galton, R. Morgan, D. Evans, C. Oliver, M. Meyer, et al. 2010. "Listening to Your Heart: How Interoception Shapes Emotion Experience and Intuitive Decision Making." *Psychological Science* 21(12), 1835–1844.

Emmons, R. A., and M. E. McCullough. 2003. "Counting Blessings Versus Burdens: An Experimental Investigation of Gratitude and Subjective Well-Being in Daily Life." *Journal of Personality and Social Psychology* 84(2), 377–389.

Eugene, A. R., and J. Masiak. 2015. "The Neuroprotective Aspects of Sleep." *Medtube Science* 3(1), 35–40.

Feinstein, B. A., R. Hershenberg, V. Bhatia, J. A. Latack, N. Meuwly, and J. Davila. 2013. "Negative Social Comparison on Facebook and Depressive Symptoms: Rumination as a Mechanism." *Psychology of Popular Media Culture* 2(3), 161–170.

Feldman, S., and N. Conforti. 1985. "Modifications of Adrenocortical Responses Following Frontal Cortex Simulation in Rats with Hypothalamic Deafferentations and Medial Forebrain Bundle Lesions." *Neuroscience* 15(4), 1045–1047.

Figueiredo, H. F., A. Bruestle, B. Bodie, C. M. Dolgas, and J. P. Herman. 2003. "The Medial Prefrontal Cortex Differentially Regulates Stress-Induced C-Fos Expression in the Forebrain Depending on Type of Stressor." *European Journal of Neuroscience* 18(8), 2357–2364.

Galynker, I. 2023. *The Suicidal Crisis: Clinical Guide to the Assessment of Imminent Suicide Risk*. 2nd ed. Oxford University Press.

Garrison, K. A., T. A. Zeffiro, D. Scheinost, R. T. Constable, and J. A. Brewer. 2015. "Meditation Leads to Reduced Default Mode Network Activity Beyond an Active Task." *Cognitive, Affective, & Behavioral Neuroscience* 15(3), 712–720.

Gerber, J. P., L. Wheeler, and J. Suls. 2018. "A Social Comparison Theory Meta-Analysis 60+ Years On." *Psychological Bulletin* 144(2), 177–197.

Gerin, W., M. J. Zawadzki, J. F. Brosschot, J. F. Thayer, N. J. S. Christenfeld, T. S. Campbell, et al. 2012. "Rumination as a Mediator of Chronic Stress Effects on Hypertension: A Causal Model." *International Journal of Hypertension* 2012, 453465.

Gianferante, D., M. V. Thoma, L. Hanlin, X. Chen, J. Breines, P. M. Zoccola, et al. 2014. "Post-Stress Rumination Predicts HPA Axis Responses to Repeated Acute Stress." *Psychoneuroendocrinology* 49, 244–252.

Gibson, L. C. 2015. *Adult Children of Emotionally Immature Parents: How to Heal from Distant, Rejecting, or Self-Involved Parents*. New Harbinger.

Goleman, D. 2005. *Emotional Intelligence: Why It Can Matter More Than IQ*. Random House Publishing Group.

Haapakoski, R., J. Mathieu, K. P. Ebmeier, H. Alenius, and M. Kivimäki. 2015. "Cumulative Meta-Analysis of Interleukins 6 and 1β, Tumour Necrosis Factor α and C-reactive Protein in Patients with Major Depressive Disorder." *Brain, Behavior, and Immunity* 49, 206–215.

Haghayegh, S., S. Khoshnevis, M. H. Smolensky, K. R. Diller, and R. J. Castriotta. 2019. "Before-Bedtime Passive Body Heating by Warm Shower or Bath to Improve Sleep: A Systematic Review and Meta-Analysis." *Sleep Medicine Reviews* 46, 124–135.

Harrison, N. A., L. Brydon, C. Walker, M. A. Gray, A. Steptoe, and H. D. Critchley. 2009. "Inflammation Causes Mood Changes Through Alterations in Subgenual Cingulate Activity and Mesolimbic Connectivity." *Biological Psychiatry* 66(5), 407–414.

Herman, J. P., J. M. McKlveen, S. Ghosal, B. Kopp, A. Wulsin, R. Makinson, et al. 2016. "Regulation of the Hypothalamic-Pituitary-Adrenocortical Stress Response." *Comprehensive Physiology* 6(2), 603–621.

Hillman, C. H., M. B. Pontifex, D. M. Castelli, N. A. Khan, L. B. Raine, M. R. Scudder, et al. 2014. "Effects of the FITKids Randomized Controlled Trial on Executive Control and Brain Function." *Pediatrics* 134(4), 1063–1071.

Horovitz, S. G., A. R. Braun, W. S. Carr, D. Picchioni, T. J. Balkan, M. Fukunaga, et al. 2009. "Decoupling of the Brain's Default Mode Network During Deep Sleep." *PNAS* 106(27), 11376–11381.

Huang, Y., A. Mohan, D. De Ridder, S. Sunaert, and S. Vanneste. 2018. "The Neural Correlates of the Unified Percept of Alcohol-Related Craving: A fMRI and EEG Study." *Scientific Reports* 8(1), 923.

Jarraya, B., P. Brugieres, N. Tani, J. Hodel, B. Grandjacques, G. Fenelon, et al. 2010. "Disruption of Cigarette Smoking Addiction After Posterior Cingulate Damage." *Journal of Neurosurgery* 113(6), 1219–1221.

Jiang, H., M. P. White, M. D. Greicius, L. C. Waelde, and D. Spiegel. 2017. "Brain Activity and Functional Connectivity Associated with Hypnosis." *Cerebral Cortex* 27(8), 4083–4093.

Kajimura, S., N. Masuda, J. K. L. Lau, and K. Murayama. 2020. "Focused Attention Meditation Changes the Boundary and Configuration of Functional Networks in the Brain." *Scientific Reports* 10(1), 18426.

Killingsworth, M. A., and D. T. Gilbert. 2010. "A Wandering Mind Is an Unhappy Mind." *Science 330*(6006), 932.

Kinney, D. K., and M. Tanaka. 2009. "An Evolutionary Hypothesis of Depression and Its Symptoms, Adaptive Value, and Risk Factors." *Journal of Nervous and Mental Disease 197*(8), 561–567.

Kuhfuß, M., T. Maldei, A. Hetmanek, and N. Baumann. 2021. "Somatic Experiencing—Effectiveness and Key Factors of a Body-Oriented Trauma Therapy: A Scoping Literature Review." *European Journal of Psychotraumatology 12*(1), 1929023.

Kuypers, L. M. 2011. *The Zones of Regulation: A Curriculum Designed to Foster Self-Regulation and Emotional Control*. Think Social Publishing.

Lanier, J. 2018. *Ten Arguments for Deleting Your Social Media Accounts Right Now*. Henry Holt and Company.

Li, W., X. Mai, and C. Liu. 2014. "The Default Mode Network and Social Understanding of Others: What Do Brain Connectivity Studies Tell Us." *Frontiers in Human Neuroscience 8*(74), 1–15.

Liston, C., B. S. McEwen, and B. J. Casey. 2009. "Psychosocial Stress Reversibly Disrupts Prefrontal Processing and Attentional Control." *PNAS 106*(3), 912–917.

Lydon-Staley, D. M., C. Kuehner, V. Zamoscik, S. Huffziger, P. Kirsch, and D. S. Bassett. 2019. "Repetitive Negative Thinking in Daily Life and Functional Connectivity Among Default Mode, Front-Parietal, and Salience Networks." *Translational Psychiatry 9*(1), 234–245.

Mayer, E. A. 2011. "Gut Feelings: The Emerging Biology of Gut-Brain Communication." *Nature Reviews Neuroscience 12*(8), 453–466.

McNaughton, D., D. Hamlin, J. McCarthy, D. Head-Reeves, and M. Schreiner. 2007. "Learning to Listen: Teaching an Active Listening Strategy to Preservice Education Professionals." *Topics on Early Childhood Special Education 27*(4), 223–231.

Miller, K. 2021, November 17. "Olympic Gold Medalist Simone Biles Uses This Simple Technique to Help Manage Her Anxiety." *Women's Health*.

Monks of New Skete and M. Goldberg. 2017. *Let Dogs Be Dogs: Understanding Canine Nature and Mastering the Art of Living with Your Dog*. Little, Brown and Company.

Morris, G. P., I. A. Clark, R. Zinn, and B. Vissel. 2013. "Microglia: A New Frontier for Synaptic Plasticity, Learning and Memory, and Neurodegenerative Disease." *Neurobiology of Learning and Memory* 105, 40–53.

Moss, A., S. Robbins, S. Achanta, L. Kuttippurathu, S. Turick, S. Nieves, et al. 2021. "A Spatially-Tracked Single Cell Transcriptomics Map of Neuronal Networks in the Intrinsic Cardiac Nervous System." *iScience* 24(7), 1–25.

Nir, Y., T. Andrillon, A. Marmelshtein, N. Suthana, C. Cirelli, G. Tononi, et al. 2017. "Selective Neuronal Lapses Precede Human Cognitive Lapses Following Sleep Deprivation." *Nature Medicine* 23(12), 1474–1480.

Nutt, D., S. Wilson, and L. Paterson. 2008. "Sleep Disorders as Core Symptoms of Depression." *Dialogues in Clinical Neuroscience*, 10(3), 329–336.

Obama, M. 2018. *Becoming*. Crown.

O'Brien, L. T., and C. S. Crandall. 2003. "Stereotype Threat and Arousal: Effects on Women's Math Performance." *Personality and Social Psychology Bulletin* 29(6), 782–789.

Ortner, N., and M. Hyman. 2014. *The Tapping Solution: A Revolutionary System for Stress-Free Living*. Hay House.

Paykel, E. S., J. K. Myers, J. J. Lindenthal, and J. Tanner. 1974. "Suicidal Feelings in the General Population: A Prevalence Study." *The British Journal of Psychiatry: The Journal of Mental Science 124*(0), 460–469.

Porges, S. W. 2021. *Polyvagal Safety: Attachment, Communication, Self-Regulation.* W. W. Norton & Company.

Radley, J. J., A. B. Rocher, W. G. M. Janssen, P. R. Hof, B. S. McEwen, and J. H. Morrison. 2005. "Reversibility of Apical Dendritic Retraction in the Rat Medial Prefrontal Cortex Following Repeated Stress." *Experimental Neurology 196*(1), 199–203.

Radley, J. J., and P. E. Sawchenko. 2011. "A Common Substrate for Prefrontal and Hippocampal Inhibition of the Neuroendocrine Stress Response." *The Journal of Neuroscience 31*(26), 9683–9695.

Raichle, M. E., A. M. MacLeod, A. Z. Snyder, W. J. Powers, D. A. Gusnard, and G. L. Shulman. 2001. "A Default Mode of Brain Function." *PNAS 98*(2), 676–682.

Ratcliffe, M. 2013. "A Bad Case of the Flu? The Comparative Phenomenology of Depression and Somatic Illness." *Journal of Consciousness Studies 21*, 198–218.

Regen, W., S. D. Kyle, C. Nissen, B. Feige, C. Baglioni, J. Hennig, et al. 2016. "Objective Sleep Disturbances Are Associated with Greater Waking Resting-State Connectivity Between the Retrosplenial Cortex/ Hippocampus and Various Nodes of the Default Mode Network." *Journal of Psychiatry and Neuroscience 41*(5), 295–303.

Restrepo, S. (Director). 2019. *Brené Brown: The Call to Courage* [Film]. Netflix.

Rogers, A., and B. Welch. 2009. "Using Standardized Clients in the Classroom: An Evaluation of a Training Module to Teach Active Listening Skills to Social Work Students." *Journal of Teaching in Social Work 29*(2), 153–168.

Savitsky, K., N. Epley, and T. Gilovich. 2001. "Do Others Judge Us as Harshly as We Think? Overestimating the Impact of Our Failures, Shortcomings, and Mishaps." *Journal of Personality and Social Psychology* 81(1), 44–56.

Schacter, D. L., D. R. Addis, and R. L. Buckner. 2007. "Remembering the Past to Imagine the Future: The Prospective Brain." *Nature Reviews Neuroscience* 8(9), 657–661.

Schaller, M., and D. R. Murray. 2008. "Pathogens, Personality, and Culture: Disease Prevalence Predicts Worldwide Variability in Sociosexuality, Extraversion, and Openness to Experience." *Journal of Personality and Social Psychology* 95, 212–221.

Schiel, J. E., F. Holub, R. Petri, J. Leerssen, S. Tamm, M. Tahmasian, et al. 2020. "Affect and Arousal in Insomnia: Through a Lens of Neuroimaging Studies." *Current Psychiatry Reports* 22(9), 44.

Schmidt, R. E., A. G. Harvey, and M. Van der Linden. 2011. "Cognitive and Affective Control in Insomnia." *Frontiers in Psychology* 2(349), 1–12.

Schrantee, A., P. J. Lucassen, J. Booij, and L. Reneman. 2018. "Serotonin Transporter Occupancy by the SSRI Citalopram Predicts Default-Mode Network Connectivity." *European Neuropsychopharmacology: The Journal of The European College of Neuropsychopharmacology* 28(10), 1173–1179.

Schwartz, R. C., and M. Sweezy. 2019. *Internal Family Systems*. 2nd ed. The Guilford Press.

Segerstrom, S. C., and G. E. Miller. 2006. "Psychological Stress and the Human Immune System: A Meta-Analytic Study of 30 Years of Inquiry." *Psychological Bulletin* 130(4), 601–630.

Sesa-Ashton, G., R. Wong, B. McCarthy, S. Datta, L. A. Henderson, T. Dawood, et al. 2022. "Stimulation of the Dorsolateral Prefrontal Cortex Modulates Muscle Sympathetic Nerve Activity and Blood Pressure in Humans." *Cerebral Cortex Communication* 3(2), tgac017.

Setiawan, E., A. A. Wilson, R. Mizrahi, P. M. Rusjan, L. Miler, G. Rajkowska, et al. 2015. "Increased Translocator Protein Distribution Volume, a Marker of Neuroinflammation, in the Brain During Major Depressive Disorder." *JAMA Psychiatry* 72(3), 268–275.

Shapiro, F. 2012. *Getting Past Your Past: Take Control of Your Life with Self-Help Techniques from EMDR Therapy*. Rodale.

Shapiro, F., and M. S. Forrest. 2004. *EMDR: The Breakthrough "Eye Movement" Therapy for Overcoming Anxiety, Stress, and Trauma*. Basic Books.

Siegel, D., and T. P. Bryson. 2012. *The Whole Brain Child: 12 Revolutionary Strategies to Nurture Your Child's Developing Mind*. Bantam.

Speed, C. 2007. "Exercise and Menstrual Function." *BMJ* 334(7586), 164–165.

Sridharan, D., D. J. Levitin, and V. Menon. 2008. "A Critical Role for the Right Front-Insular Cortex in Switching Between Central-Executive and Default-Mode Networks." *PNAS* 105(34), 12569–12574.

Stern, Y., A. MacKay-Brandt, S. Lee, P. McKinley, K. McIntyre, Q. Razlighi, et al. 2019. "Effect of Aerobic Exercise on Cognition in Younger Adults." *Neurology* 93(4), 185.

Suh, S., H. Kim, T. T. Dang-Vu, E. Joo, and C. Shin. 2016. "Cortical Thinning and Altered Cortico-Cortical Structural Covariance of the Default Mode Network in Patients with Persistent Insomnia Symptoms." *Sleep* 39(1), 161–171.

Sullivan, R. M., and A. Gratton. 1999. "Lateralized Effects of Medial Prefrontal Cortex Lesions on Neuroendocrine and Autonomic Stress Responses in Rats." *The Journal of Neuroscience* 19(7), 2834–2840.

Tamaki, M., J. W. Bang, T. Watanabe, and Y. Sasake. 2016. "Night Watch in One Brain Hemisphere During Sleep Associated with the First-Night Effect in Humans." *Current Biology* 26(9), 1190–1194.

Tanabe, J., E. Nyberg, L. F. Martin, J. Martin, D. Cordes, E. Kronberg, et al. 2011. "Nicotine Effects on Default Mode Network During Resting State." *Psychopharmacology* 216(2), 287–295.

Thomas, M. L., H. C. Sing, G. Belenky, H. H. Holcomb, H. S. Mayberg, R. F. Dannals, et al. 2003. "Neural Basis of Alertness and Cognitive Performance Impairments During Sleepiness: II. Effects of 48 and 72 h of Sleep Deprivation on Waking Human Regional Brain Activity." *Thalamus & Related Systems* 2(3), 199–299.

Thomason, M. E., J. P. Hamilton, and I. H. Gotlib. 2011. "Stress-Induced Activation of the HPA Axis Predicts Connectivity Between Subgenual Cingulate and Salience Network During Rest in Adolescents." *Journal of Child Psychology & Psychiatry* 52(10), 1026–1034.

van der Kolk, B. 2014. *The Body Keeps the Score: Brain, Mind, and Body in the Healing of Trauma*. Penguin Books.

Verduyn, P., D. S. Lee, J. Park, H. Shablack, A. Orvell, J. Bayer, et al. 2015. "Passive Facebook Usage Undermines Affective Well-Being: Experimental and Longitudinal Evidence." *Journal of Experimental Psychology: General* 144(2), 480–488.

Weger, H., G. C. Bell, E. M. Minei, and M. C. Robinson. 2014. "The Relative Effectiveness of Active Listening in Initial Interactions." *The International Journal of Listening* 28(1), 13–31.

Wu, J. C., J. C. Gillin, M. S. Buchsbaum, P. Chen, D. B. Keator, N. K. Wu, et al. 2006. "Frontal Lobe Metabolic Decreases with Sleep Deprivation Not Totally Reversed by Recovery Sleep." *Neuropsychopharmacology* 31(12), 2783–2792.

Xie, L., H. Kang, Q. Xu, M. J. Chen, Y. Liao, M. Thiyagarajan, et al. 2013. "Sleep Drives Metabolite Clearance from the Adult Brain." *Science* 342(6156), 373–377.

Xiong, H., R. J. Guo, and H. W. Shi. 2020. "Altered Default Mode Network and Salience Network Functional Connectivity in Patients with Generalized Anxiety Disorders: An ICA-based Resting-State fMRI Study." *Evidence-Based Complementary and Alternative Medicine* 2020, 4048916.

Zheng, H., L. Kong, L. Chen, H. Zhang, and W. Zheng. 2015. "Acute Effects of Alcohol on the Human Brain: A Resting-State fMRI Study." *BioMed Research International 2015*, 947529.

Zhou, H., X. Chen, Y. Shen, L. Li, N. Chen, Z. Zhu, et al. 2020. "Rumination and the Default Mode Network: Meta-analysis of Brain Imaging Studies and Implications for Depression." *NeuroImage* 206, 116287.

Betsy Holmberg, PhD, is a psychologist and author specializing in overthinking and negative self-talk. She writes for *Psychology Today*, and has been featured on radio, television, and podcasts. Before that, she ran the mental health service line at McKinsey & Company, and received her PhD from Duke University.

Real change *is* possible

For more than fifty years, New Harbinger has published proven-effective self-help books and pioneering workbooks to help readers of all ages and backgrounds improve mental health and well-being, and achieve lasting personal growth. In addition, our spirituality books offer profound guidance for deepening awareness and cultivating healing, self-discovery, and fulfillment.

Founded by psychologist Matthew McKay and Patrick Fanning, New Harbinger is proud to be an independent, employee-owned company. Our books reflect our core values of integrity, innovation, commitment, sustainability, compassion, and trust. Written by leaders in the field and recommended by therapists worldwide, New Harbinger books are practical, accessible, and provide real tools for real change.

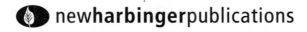

MORE BOOKS from
NEW HARBINGER PUBLICATIONS

THE INNER CRITIC WORKBOOK

Self-Compassion and Mindfulness Skills to Reduce Feelings of Shame, Build Self-Worth, and Improve Your Life and Relationships

978-1648484292 / US $25.95

RELEASING TOXIC ANGER FOR WOMEN

Somatic Practices and CBT Skills to Transform Negative Thoughts, Soothe Stress, and Stay True to Yourself

978-1648483295 / US $19.95

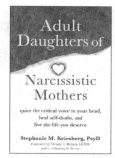

ADULT DAUGHTERS OF NARCISSISTIC MOTHERS

Quiet the Critical Voice in Your Head, Heal Self-Doubt, and Live the Life You Deserve

978-1648480096 / US $18.95

THE EMOTIONALLY EXHAUSTED WOMAN

Why You're Feeling Depleted and How to Get What You Need

978-1648480157 / US $20.95

SIMPLE WAYS TO UNWIND WITHOUT ALCOHOL

50 Tips to Drink Less and Enjoy More

978-1648482342 / US $18.95

TOXIC STRIVING

Why Hustle and Wellness Culture Are Leaving Us Anxious, Stressed, and Burned Out—and How to Break Free

978-1648484063 / US $19.95

newharbingerpublications

1-800-748-6273 / newharbinger.com

(VISA, MC, AMEX / prices subject to change without notice)

Follow Us

Don't miss out on new books from New Harbinger.
Subscribe to our email list at **newharbinger.com/subscribe**

Did you know there are **free tools** you can download for this book?

Free tools are things like **worksheets, guided meditation exercises,** and **more** that will help you get the most out of your book.

You can download free tools for this book—whether you bought or borrowed it, in any format, from any source—from the New Harbinger website. All you need is a NewHarbinger.com account. Just use the URL provided in this book to view the free tools that are available for it. Then, click on the "download" button for the free tool you want, and follow the prompts that appear to log in to your NewHarbinger.com account and download the material.

You can also save the free tools for this book to your **Free Tools Library** so you can access them again anytime, just by logging in to your account! Just look for this button on the book's free tools page. ⟶ `+ Save this to my free tools library`

If you need help accessing or downloading free tools, visit **newharbinger.com/faq** or contact us at **customerservice@newharbinger.com**.